# IMPROVING EQUITY
# IN DATA SCIENCE

*Improving Equity in Data Science* offers a comprehensive look at the ways in which data science can be conceptualized and engaged more equitably within the K-16 classroom setting, moving beyond merely broadening participation in educational opportunities. This book makes the case for field wide definitions, literacies, and practices for data science teaching and learning that can be commonly discussed and used, and provides examples from research of these practices and literacies in action.

Authors share stories and examples of research wherein data science advances equity and empowerment through the critical examination of social, educational, and political topics. In the first half of the book, readers will learn how data science can deliberately be embedded within K-12 spaces to empower students to use it to identify and address inequity. The latter half will focus on equity of access to data science learning opportunities in higher education, with a final synthesis of lessons learned and presentation of a 360-degree framework that links access, curriculum, and pedagogy as multiple facets collectively essential to comprehensive data science equity work.

Practitioners and teacher educators will be able to answer the question, "how can data science serve to move equity efforts in computing beyond basic inclusion to empowerment?" whether the goal is to simply improve definitions and approaches to research on data science or support teachers of data science in creating more equitable and inclusive environments within their classrooms.

**Colby Tofel-Grehl** is an associate professor of STEM teacher education and learning at Utah State University, USA.

**Emmanuel Schanzer** is a math and CS-education researcher, and the co-founder and chief curriculum architect at Bootstrap.

# IMPROVING EQUITY IN DATA SCIENCE

## Re-Imagining the Teaching and Learning of Data in K-16 Classrooms

*Edited by Colby Tofel-Grehl
and Emmanuel Schanzer*

Routledge
Taylor & Francis Group

NEW YORK AND LONDON

Designed cover image: © Getty Images

First published 2024
by Routledge
605 Third Avenue, New York, NY 10158

and by Routledge
4 Park Square, Milton Park, Abingdon, Oxon, OX14 4RN

*Routledge is an imprint of the Taylor & Francis Group, an informa business*

ISBN: 978-1-032-42866-6 (hbk)
ISBN: 978-1-032-42862-8 (pbk)
ISBN: 978-1-003-36463-4 (ebk)

DOI: 10.4324/9781003364634

Typeset in Times New Roman
by KnowledgeWorks Global Ltd.

# CONTENTS

# AUTHOR BIOS

## Editors

Dr. Emmanuel Schanzer spent several years as a program manager and developer before becoming a public high school teacher and middle school academic coach in Boston. He is the co-founder and Chief Curriculum Architect at Bootstrap, which he first designed as a curriculum for his own students. Emmanuel has long been involved in connecting educators and technology, connecting parties at NCTM, CSTA, Google, Microsoft, Facebook and at universities across the country. He holds degrees in computer science and curriculum development and completed his doctoral studies at Harvard with a research focus on using programming to teach algebra.

Dr. Colby Tofel-Grehl is an associate professor in the School of Teacher Education & Leadership at Utah State University. Her scholarship interrogates the structures, systems, and practices that foster inequities across STEM learning environments. She designs teacher professional development and curricular materials within the frameworks of rightful presence and consequential learning to facilitate minoritized youth's success within STEM spaces. Her work leverages the affordances of making and data science to center students' identities and cultures within authentic and empowering STEM work. She received the 2019 Award for Significant Contribution to Educational Measurement and Research Methodology from Division D of the American Educational Research Association and the 2020 Early Career Science Teacher Educator of the Year Award from the Association for Science Teacher Educators. In 2021, the National Science Foundation's ITEST Program recognized her grant, Project ESTITCH, as one of three nationally outstanding projects in the service of broadening participation in STEM.

# CONTRIBUTORS

**June Ahn** is a Professor of Learning Sciences and Research-Practice Partnerships at UC Irvine. He conducts research at the intersection of participatory design, technology, education, and community partnerships. His research supports the development of research practice partnerships for supporting technology and educational ecosystems.

**Amanda Barany** is a postdoctoral scholar at the University of Pennsylvania Graduate School of Education, working in the Penn Center for Learning Analytics. She is a Co-PI for the Coding Like a Data Miner Project, a data science-based computer science curriculum that uses culturally relevant and responsive pedagogies to center diverse learners in STEM. Other areas of work involve leveraging quantitative ethnographic techniques to examine complex learning processes. She completed her graduate work at Drexel University's Games and Learning in Interactive Digital Environments lab with a focus on game-based learning, computer-based learning environments, identity, and motivation.

**Melissa (Missy) Barker** is a Ph.D. student at Utah State University studying Curriculum and Instruction-Mathematics emphasis. She received her B.S. degree in Electronics Engineering Technology from Brigham Young University and Master of Mathematics from USU. She currently teaches high school mathematics, concurrent enrollment, and an undergraduate course in mathematics teaching for USU. Her research interests lie in promoting STEM interest through project-based learning opportunities.

**Tiffany Barnes** is a Distinguished Professor of Computer Science at North Carolina State University and Founding Co-Director of the STARS Computing Corps, a Broadening Participation in Computing Alliance funded by the National Science Foundation. She is a Distinguished Member of the Association of Computing Machinery (ACM). Her internationally recognized research program focuses on transforming education with AI-driven learning technologies and research on equity and broadening participation. Her current research ranges from investigations of intelligent tutoring systems and teacher professional development to foundational work on educational data mining, computational models of interactive problem-solving, and design of computational thinking curricula. Her personalized learning technologies and broadening participation programs have impacted thousands of K-20 students throughout the United States.

**Alan Barrera** is a doctoral student in Education at the University of Texas at El Paso. He is a STEM researcher in the fields of bioengineering and computer science. His work includes the development of the Coding Like a Data Miner curriculum. Other work includes teaching a workshop consisting of an introduction to Biomaking, and another one using biomaterials as ways for making and learning. Other studies include the analysis of discourse in social media and a study that establishes opportunities to learn about Biomaking. He also worked as a teaching assistantship in STEM doctoral courses.

**Brian Broll** is a Research Scientist at the Institute for Software Integrated Systems at Vanderbilt University. He holds Ph.D. in Computer Science from Vanderbilt University and a B.Sc. degree in Mathematics Education from Buena Vista University. His research interests include computer science education, model-integrated computing, along with applied artificial intelligence and machine learning. Brian Broll is also the Chief Architect of NetsBlox, an educational block-based programming environment designed for teaching advanced concepts to high school students.

**Jessica Cai** is a master's student at Harvard Graduate School of Education. During her undergraduate education at the University of California, Irvine, she engaged as a research assistant in the Design and Partnership Lab.

**Virginia Catété** specializes in research bringing advanced technologies including artificial intelligence, game design, and security to middle grades audiences. Currently serving as Assistant Professor of Computer Science at North Carolina State University, her research spans student scaffolding, curriculum creation, and teacher professional development. With a Ph.D. in Computer Science, Dr. Catété publishes extensively in top-tier Computing

Education conferences and journals, winning multiple best paper awards and advancing knowledge in Computing Education Research. Dr. Cateté has also led 15 years of K-12 computer science and engineering outreach, both across the United States and overseas in Rwanda, impacting thousands of students annually.

**David Feldon** is a Professor of Instructional Technology and Learning Sciences in the College of Education and Human Services at Utah State University. From a disciplinary lens of educational psychology, his scholarship identifies mechanisms of learning and education that facilitate the equitable development of expertise—specifically in STEM disciplines. To understand the mechanisms that shape individual learning, his work also engages the sociological factors that drive both experiences and opportunities. His research builds bridges from a deep understanding of motivation and cognition to broader cultural and structural influences that shape divergent educational pathways and various modes of success.

**Isabella Gransbury** is a Ph.D. candidate in the department of Computer Science at North Carolina State University in Raleigh, NC. Her research is centered around creating equitable learning environments for novice and experienced computer science students. She has great interest in improving professional developments and training for K-12 computer science teachers.

**Shuchi Grover** is a Learning and Computer Scientist and STEM Education Researcher. Her research centers on computational thinking, CS education and STEM learning in K-14 spaces. She holds a Ph.D. in Learning Sciences and Technology from Stanford University, master's degrees in Education (Harvard University) and computer science in addition to bachelor's degree in computer science and physics.

**Mark Guzdial** is a Professor in Computer Science & Engineering and the Director of the Program in Computing in the Arts and Sciences at the University of Michigan. He was one of the founders of the International Computing Education Research conference. He is an ACM Distinguished Educator and a Fellow of the ACM. When not shaping the field of computer science education, Mark can be found playing ukulele much to the joy of his colleagues and friends.

**Jenny Han** is a member of the UCI Design and Partnership Lab. She graduated from Stanford with masters in Human Computer interaction focused on learning technologies. Her work focuses on ways to make computer science education more equitable and personally meaningful for all.

**Tyler Hansen** (he/him) is a former high school Science Teacher and recent recipient of Utah's Biology Teacher of the year. A Ph.D. student at Utah State University, his goal is to bring more authentic science experiences in the K-12 classroom settings. His research interests include scientific literacy, technology in the science classroom, and biology education research. When he's not working, Tyler enjoys the great outdoors by backpacking, biking, and trying to identify as many species as possible along the way.

**Devin Jean** is a graduate student at Vanderbilt University in computer science. He graduated with a B.S. in computer science from Middle Tennessee State University in 2020. He works as a Research Assistant focused on developing educational software.

**Mengying Jiang** is a Ph.D. student in Instructional Technology and Learning Sciences at Utah State University. Her research interests focus on design-based research, culturally responsive computing, and learning environment design.

**Michael A. Johnson** is a doctoral student at the University of North Texas in Learning Technologies, who combines over a decade of experience as a public school teacher in computer science, business, and technology with his research in learning sciences, instructional design, and computing education. His work encompasses digital literacy, computer and data science education, and the impacts of computing. As a research assistant, he contributed to the innovative "Coding Like a Data Miner" curriculum project, focusing on participatory curriculum co-design, computational thinking, and culturally relevant pedagogy for pre-college data science education.

**Akos Ledeczi** is a Computer Science Professor and a founding member of the Learning Incubator: a Vanderbilt Initiative (LIVE) at Vanderbilt University. His current research focuses on introductory computer science education in K-12. His group has developed an open source web-based visual programming environment called NetsBlox that was specifically designed to introduce cutting edge computing concepts, such as distributed computing, networking, the Internet of Things, cybersecurity and machine learning, to novices. The environment and associated curriculum are being used in a number of schools around the United States. Dr. Ledeczi has co-authored multiple massive open online courses, one of which called Introduction to Programming with MATLAB has been taken by well over 300,000 students from 190+ countries.

**Victor R. Lee** is an Associate Professor in the Stanford Graduate School of Education. His research involves the study and design of STEM learning experiences with an emphasis on supporting teaching and learning with data, AI literacy, elementary computer science education, and science teaching and learning. He is past recipient of an NSF CAREER Award, AERA's Jan Hawkins Award, and a National Academy of Education/Spencer Foundation Postdoctoral

Fellowship. He is a member of the editorial boards of several leading journals and is past president and an elected fellow of the International Society of the Learning Sciences.

**Christopher Martinez** is Lab Coordinator for the University of California, Irvine's Design and Partnership Lab. His work focuses on designing educational ecosystems that improve interest and engagement.

**Ha Nguyen** is an Assistant Professor of Instructional Technology and Learning Sciences at Utah State University. Dr. Nguyen's research integrates learning sciences, learning analytics, and human-centered design to promote deeper learning in Science, Technology, Engineering, and Math (STEM) contexts for diverse learners. Her research explores how to give students the opportunity to collaborate and reflect on their learning through design partnerships and analytics.

**Sayed Mohsin Reza**, currently serving as an Assistant Professor at Pennsylvania State University Harrisburg, is a researcher in the fields of Data Science, Machine Learning, and Software Engineering. His scholarly endeavors involve the development of innovative data science algorithms and techniques aimed at enhancing data-driven decision-making processes. Reza is engaged in the development of a Data Science based CS curriculum for pre-college learners. Noteworthy contributions include leveraging data science for advanced software analysis for Nuba Solutions, spearheading data-driven decision-making in peer-to-peer review management, PROCONF, and developing a data science-based CS curriculum for an NSF project, CLDM. Recognizing his contributions, Reza was honored with the prestigious 2023 IEEE FIE New Faculty Fellow Award.

**Fernando Rodriguez** is an Assistant Professor of teaching at UC Irivine's School of Education. His research focuses on using learning analytics to better understand student achievement in technology-enhanced and online STEM courses.

**Kristin A. Searle** is an Associate Professor of Instructional Technology and Learning Sciences at Utah State University. She received her Ph.D. in education and anthropology from the University of Pennsylvania. Her work focuses on how participating in making activities can broaden students' sense of what computing is and who can do it, with a focus on the development of culturally responsive computing pedagogies. Her work has appeared in journals such as *Harvard Educational Review* and the *British Journal of Educational Technology*.

**Tamara L. Shreiner** is Associate Professor in the History Department at Grand Valley State University in Allendale, Michigan, where she specializes in social studies education. She taught middle and high school history, geography, and civics for nearly ten years. Her research focuses on disciplinary literacy in social studies, particularly the role of data literacy in supporting disciplinary literacy. Among her publications are articles in *Cognition and Instruction, Theory & Research in Social Education*, the *Journal of Social Studies Research, Social Education*, and the *British Journal of Educational Technology*.

**Emily Slater** is a doctoral student in Instructional Technology and Learning Sciences at Utah State University (USU). Her previous studies include a Bachelor of Sociology from the University of the Fraser Valley, and Master of Public Administration from the University of Victoria—both of which are in her home-province of British Columbia, Canada. Emily does research and instructs at USU where she is continually inspired by the community around here. She is active in student leadership with the Instructional Technology Student Association and the USU Intersections Center.

**Mario Suárez** is an Assistant Professor at Utah State University. He is interested in the intersection of race, gender identity, sexuality, and class, and how binary notions of gender permeate K-12 STEM spaces and curricula. His intersectional research agenda broadly asks: How does our understanding of gender and sexuality shape K-12 education (e.g., standards, curriculum, bathrooms, group placements, sports, STEM, teachers and teaching)? He tries to answer this question through a primarily critical quantitative lens using nationally representative data, though he also draws on qualitative methods when large data is not available. His research in education draws from the fields of sociology, demography, gender studies, and policy studies.

**Seth Van Doren** is a PhD Student in UC Irvine School of Education's TLEI Program. His scholarship focuses on science learning and participatory design research methodologies. He received the B.A. degree in chemistry (with a minor STEM education) from the University of California Berkeley, Berkeley, CA, USA, in 2019. His research interests include authentic STEM investigations in K-12 settings, social learning, and increasing student epistemic agency.

**Justice Toshiba Walker** is a Learning Scientist and Assistant Professor at the University of Texas at El Paso, where he leads the ABC Learning Lab and its emphasis on learning technologies that leverage next generation bio and computing tools. His research examines how middle and high school youth respond to learning paradigms that emphasize cultural relevance, epistemological agency and critical literacies. Walker holds a B.S. degree in Molecular Biology and English Literature from the University of Miami (FL), an M.S. degree in Biotech Engineering and Ph.D. in Teaching and Learning, both earned at the University of Pennsylvania (UPenn).

# FOREWORD

> In a child's power to master the multiplication table there is more sanctity than in all your shouted 'Amens!', 'Holy, Holies!' and 'Hosannahs!' An idea is a greater monument than a cathedral. And the advance of man's knowledge is more of a miracle than any sticks turned to snakes, or the parting of waters!
>
> – Henry Drummond, Inherit the Wind

Jerome Lawrence and Robert Lee could not have imagined, when they penned Inherit the Wind in 1955, that their retelling of the 1925 Scope's Monkey Trial would withstand changing times and issues to remain relevant and insightful nearly 70 years later. Today, public education and public educators find themselves set between angry lawmakers who seek to protect students from allegedly dangerous ideas, and knowledge their students need to become engaged, informed citizens.

As former teachers, we have watched the current shift towards data science education with interest and caution. While data science education holds transformative potential to increase the deep thinking skills we believe sit at the heart of meaningful and consequential (Hall and Jurow, 2015) learning, we also recognize that in every school across the country, teachers feel overwhelmed by the never-ending list of additions to their curricula and their responsibilities. Furthermore, how will data science be engaged and enacted in these overpacked, standards driven classrooms? Will it enhance the dialogues already happening, foster new ones, and provide an entryway to topics from computing to civics that is grounded in data? Or will it be siloed, flattened into another box to be checked, and hollowed out into the husk of another thinly-veiled class on quantitative thinking without context?

In thinking about the need for robust data science education within schools, we returned to the purpose of free public education in the United States. Public education in the United States serves as possibly the greatest public works endeavor undertaken in our nation's history. As noted:

> The Founding Fathers maintained that the success of the fragile American democracy would depend on the competency of its citizens. They believed strongly that preserving democracy would require an educated population that could understand political and social issues and would participate in civic life, vote wisely, protect their rights and freedoms, and resist tyrants and demagogues.

Today, as political and lobbying groups work to undercut and devalue public education, demagogues abound. Divisive politics are our norm. We see book banning, state laws prohibiting teachers from acknowledging the existence of LGBTQ+ folks, and revisionist histories where students could be taught that enslavement was good for Black Americans. The trend toward codifying misinformation into free public education is terrifying. As educators, we return to data science education as a space that can push back against the tidal surge of demagoguery within public education spaces.

Learning about and engaging with data will be necessary for future careers and engaged citizenship. But data and data science can be a double-edged sword. While it can elucidate, it can also be bent and twisted to obfuscate and misrepresent. Thus, the double-edged sword of data serves as the impetus for this volume. We believe that while data science education is necessary for all students, in and of itself, it is not sufficient. A great math class connects its content to the real world, pre-empting the question "when am I ever going to use this?" We learn to understand mathematical formulas, not because they are important in and of themselves, but because they allow us to reason about and challenge the world around us. For data science education to serve its purpose within public education, both at the elementary and secondary levels, it must be rigorous and equity focused. It is not sufficient for student to examine data; they must also examine the sources of the data, positionality of the data scientist, and missing data (proximally and distally) that might reveal a more nuanced story. Just as math education must be about more than formulas, data science education must be about more than data.

This volume is intended to share beginning thoughts and examples of what robust data science education can look like. It is by no means a definitive volume. As editors, we have trusted chapter authors to showcase the stories of their data as they believed truest to the data they worked with. Authors have shared positionality statements to help center and explicate their approach and personhood within their work. Far too often, the work of equity and justice is so often left to be shouldered by people of color. Doing so makes engaging the hard conversations added work for those historically marginalized. As two white people, both with professional academic histories, we struggled with the question of whether we were the right folks to edit this volume. Ultimately, we decided to engage this volume with the intent of starting a conversation, not creating a definitive approach. We conceptualized the role of editors as facilitators of the work, not tailors of authors messages or perceptions. As former classroom teachers we felt strongly that practitioner-centered thinking was also essential for such a work and that these perspectives were often absented from scholarly volumes. We hope this book supports, uplifts, and showcases the many ways that educators are engaging data science education in classrooms to support robust critical thinking and nuanced dialogue in our learning spaces.

## References

Hall, R., & Jurow, A. S. (2015). Changing concepts in activity: Descriptive and design studies of consequential learning in conceptual practices. *Educational Psychologist*, *50*(3), 173–189.

Kober, N., & Rentner, D. S. (2020). *History and evolution of public education in the US*. Center on Education Policy.

Lawrence, J., & Lee, R. E. (2000). *Inherit the wind*. Dramatists Play Service Inc.

# 1

# OVERVIEW

*Emmanuel Schanzer and Colby Tofel-Grehl*

## Why Data Science?

Educators have a lot on their plates. Your average teacher, administrator, or curriculum coordinator spends their days balancing the endless pressure to improve student's achievement across a variety of subjects, the need to provide physical and mental health services to students who walk in the door with any number of challenges, and the conflicting calls for more reform, more social justice, more computer science, and more focus on "the basics". States, schools, and districts are slowly reopening from the pandemic, and they are still in the throes of a well-funded and widely supported effort to get "Computer Science for All" to students. By all accounts, this is a difficult time to argue in favor of more disruptive reforms.

Yet, calls for schools everywhere to teach data science are growing. While many organizations have been quietly working on data science initiatives for years, the national conversation around data science as a must-have reform is relatively new. In a Freakonomics podcast in August 2019, Steve Levitt called for a shakeup to the traditional math curriculum that would create a home for data science. Those in industry joined the chorus, arguing for the importance of including these skills in K-12 schools and offering funding to develop curricula. By 2021, the U.S. Department of Education was hosting webinars on the subject, and funding agencies were soliciting proposals that embraced data science education.

With everything schools are juggling right now, this seems like a difficult time to add data science to the mix. The need for data science isn't going away, so where is the sense of urgency coming from in the minds of so many stakeholders? Three of the biggest reasons for this push fall into the umbrella categories of Jobs, Social Impact, and Math Outcomes.

DOI: 10.4324/9781003364634-1

*Equitable Employment*

Education is often billed as a ticket to financial stability and future success, especially with families in lower-income communities. And while getting a job cannot and should not be the only reason for school to exist, at least *one* role of the American educational system is to help students start their careers with the highest income and the most stability and security possible in a way that ensures equitable access to these opportunities.

In mature fields where the supply of skilled employees is in balance with the demand, paid internships are nearly impossible to find – even after college. The high-paying jobs in these fields require graduate degrees, or years of (mostly unpaid) experience. This creates roadblocks for students with less-affluent backgrounds because "getting the right degree" simply isn't enough to have their efforts pay off. Students who can afford the risk of taking on loans for graduate school or spending a few years building their resume are better positioned to fill those high-paying positions. And for those unable to take on massive loans or accumulate years of unpaid experience, those high-paying jobs are often out of reach.

For the moment, at least, data science is different. In 2017, IBM published a report entitled "The Quant Crunch: How the Demand for Data Science Skills Is Disrupting the Job Market." The report argued that the need for Data Scientists was already vastly outstripping supply and would only continue to grow in many sectors, creating an inflection point that would change the dynamics of the job market. In 2022, "Data Scientist" is still one of the top three positions in the United States, with a high job satisfaction and a six-figure median starting salary (from Glassdoor's, n.d., "50 Best Jobs in America").

This has created a market for data science "bootcamps", of which some promise 100% employment after just a few *weeks* of training. Regardless of whether or not these bootcamps live up to the claim, their mere existence speaks to the shockingly high demand for employees with even basic skills in data science. The amount of training required to get students started with an internship is vastly smaller for data science than for fields like law, medicine, or computer science (CS). Imagine, a situation where a high school senior had the skills to get a high-paying internship before college. This asymmetry between supply and demand will inevitably level out, with certifications and degrees creating a more formal pipeline for those with the privilege to participate. But for now, data science presents a unique moment in time where the playing field of graduation-to-high-paying-job seems more level than it has in years.

*Social Impact*

Of course, schools are about far more than merely preparing students for the workforce. Since their inception, public schools in the United States have been implicitly – and some explicitly – tasked with preserving democracy itself. Thomas Jefferson wrote that "an educated citizenry is a vital requisite for our survival as a free people". In writing about the role of education in preserving democracy, George Washington argued "To the security of a free Constitution it [Education] contributes in various ways: By convincing those, who are entrusted with the public administration, that every valuable end of Government is best answered by the enlightened confidence of the people: And by teaching the people themselves to know and to value their own rights; to discern and provide against invasions of them; to distinguish between oppression and the necessary exercise of lawful authority".

Understanding data science has become increasingly important to understanding how the government functions and how it impacts citizens. An ever-growing number of laws, policies, and public projects are informed by data science. How do we determine if a law is biased for or against a particular group of people? How do we detect whether an unexpected outcome in a district was due to voter fraud? How do we draw those voting districts, estimate growth, or model disease epidemics? As data science becomes increasingly useful in local, state, and federal decision-making, the definition of "an educated citizenry" has shifted to include an awareness of data, where it comes from, and how it can be used. If we want students to think critically about what they see and read, we must be prepared to teach them how to critique a study, identify threats to validity, or ask follow-up questions about sourcing, bias, and analysis.

Data science has also become a critical element when it comes to safety and security. At a fundamental level, we expect K-12 schools to teach our children how to protect themselves and the society they inhabit. We expect them to protect their health by teaching them about biology, nutrition, and exercise. We expect schools to protect our children from the dangers of drugs, gangs, and strangers. We expect them to teach our children about the atrocities of war and the horrors of authoritarianism, bigotry, and hatred. Many schools already teach our children about "being safe on the internet", motivated by this same expectation. But modern internet safety and privacy is about much more than teaching students not to share their passwords.

The use of sophisticated data mining and data aggregation has exploded in recent years. The advertisements we see on the websites we visit are based on what someone else has gathered about our gender, race, level of income and education, location, and more. Seemingly innocent surveys that ask "what was your first car?" or "click on your horoscope sign to see which superhero is most like you" are actually massive data trawling systems, vacuuming up clues about our age and birthday. The reason these efforts work is that each data point harvested is innocuous on its own, and most people lack data science awareness to understand how these points become problematic when they are aggregated.

If we want students to be careful about the data they share, we must be prepared to teach them how it can be used.

### Mathematical Rigor

Given the central role that mathematics plays in a balanced data science class, many see the current drive for data science education as a vehicle to address persistent gaps in statistics education. The GAISE-II report, released at the end of 2020, lays out a framework for data science education with an emphasis on statistics and mathematics. A foundation in statistics is becoming essential in many fields, and even the National Research Council (2013) includes a set of statistics standards. Despite the rise in demand for students with a statistics background, students' pre-college exposure to statistics remains anemic at best. Some elements of statistics are spread across multiple grade levels, seemingly tucked in wherever they might fit. Algebra and geometry, for example, are required classes offered by practically 100% of high schools in the United States and are taken by nearly every single student. One in four U.S. high schools, however, doesn't offer any statistics class at all – and less than one in four students (23%) takes a stats class (Loews, 2016) before college. And those who do take stats tend to be wealthier than those who don't, with 26% being higher-income students, compared to just 19% of lower-income students.

In contrast, there are some who see data science primarily as a *programming class*, where the focus lies on writing code that consumes, transforms, and displays data. Teaching data science without a rigorous statistical foundation is like a cooking class that focuses on learning how to use the equipment: students will know how to use the tools but won't be able to understand whether the end result makes sense or is accurate in any way.

### What Is Data Science?

The lack of clear terminology around data science presents serious challenges in K-12 education. In the K-12 education space, we now see movements toward "Data Literacy", "Data Acumen", "Data Awareness", and "Data Skills" marketed as being somehow different from but related to data science. Sometimes these terms are related to substantive differences in content, but new terms can also be created for strictly political ends. Schools that are unable or unwilling to offer data science may abuse the vagueness of the term to slap a new label on an existing class, or to claim that *all* of their classes teach "Data Skills". Researchers attempting to compare the impact of data science courses on students face an uphill battle when the content and goals of each course differ radically from the other. Funders who wish to support data science initiatives struggle to quantify the impact of their investment, and curriculum designers need clear definitions on which to build. Clear definitions are also critical to equity: the difference between the haves and the have-nots is often swept under the rug of vague terminology.

In this book, we have made a concerted effort to use a clear set of vocabulary throughout each chapter, drawing a simple distinction between "Data Science" and "Data Literacy". It is our hope that by focusing on cohesive and clear definitions that we can support educators in 1) understanding what data science is and 2) understanding how it can be engaged within their classrooms.

### Data Science

Broadly speaking, data science is the work of analyzing a dataset to examine a question. Within that work lies a subset of skills and tenets that drive this analysis, which are a synthesis of computing, scientific, mathematical, and statistical processes brought to bear on the intellectual pursuit of examining data.

Data science in K-12 can be thought of as – at a minimum – a synthesis of computing and statistical content. It sits at the intersection of these fields with the unique goal of furthering knowledge and understanding through the processing and examination of data. This synthesis must happen at a deep level: simply gluing a semester of computing to a semester of statistics will not result in a full-year data science course. Individually, these components have a host of procedural skills and concepts: a statistics class is more than merely memorizing formulas, and a computer science class is more than memorizing algorithms. Knowing that there are multiple sampling mechanisms in statistics is a start, but students must learn the difference between these mechanisms and when it is appropriate to use them. Knowing that there are different computational mechanisms to perform an analysis is good, but students must learn how these mechanisms can affect the accuracy of an analysis and when to use them. Finally, these skills and concepts must be kept in balance. A great computer science class that tacks on some statistics at the end isn't data science, and neither is a great math class that throws in a few weeks of coding. In a data science curriculum, the whole must be greater than the sum of its parts.

But is data science a *computing* class with some statistics, or a *statistics* class with some computing? The answer often boils down to who is asked (a source of tension we will address shortly!). One of the most popular definitions of data science is the GAISE-II report, released in November 2020 as part of a collaboration between the American Statistical Association (ASA) and the National

Council of Teachers of Mathematics (NCTM). The report is also the foundation of YouCubed's "Big Ideas in Data Science", which is no surprise given the math focus of the committee behind GAISE-II. The four Big Ideas are:

1 **Formulate statistical investigative questions** – Students generate ideas and ask questions, creating and refining statistical investigative questions.
2 **Collect/consider data** – Students learn what counts as data (e.g., visuals, sounds, numbers, categories) and understand that people collect data to answer questions. Students develop strategies to collect and organize data of various types and from various sources. Students design studies to answer statistical investigative questions.
3 **Analyze data** – Students develop ways to represent and interrogate data to notice, describe, and analyze patterns. Students recognize variability and use technology to develop models that incorporate statistical measures.
4 **Interpret and communicate** – Students decide key results to include in a data report that answers the statistical investigative question. Students communicate their results through, for example, a data visual, a poster, a video, and a data story. Students explore and share explanations, paying careful attention to what conclusions the data supports. They consider which alternatives are reasonable given the variability in findings.

This framing puts statistics and mathematical thinking at the heart of data science. The computational element of data science is almost completely absent from the Big Ideas, save for the last phrase in Big Idea 3: "use technology to develop models that incorporate statistical measures".

Those in the computing world offer a different perspective, which is similarly biased toward computing rather than statistics. These definitions put tools and computing at the heart of data science. Math and statistics are barely mentioned, and instead the definitions are filled with use-cases, computing tools, and devices. Consider IBM's list of activities performed by a Data Scientist:

- Apply mathematics, statistics, and the scientific method
- Use a wide range of tools and techniques for evaluating and preparing data – everything from SQL to data mining to data integration methods
- Extract insights from data using predictive analytics and artificial intelligence (AI), including machine learning and deep learning models
- Write applications that automate data processing and calculations
- Tell – and illustrate – stories that clearly convey the meaning of results to decision-makers and stakeholders at every level of technical knowledge and understanding
- Explain how these results can be used to solve business problems

...or Oracle's definition of data science:

Data science combines multiple fields, including statistics, scientific methods, artificial intelligence (AI),

and data analysis, to extract value from data. Those who practice data science are called data scientists,

and they combine a range of skills to analyze data collected from the web, smartphones, customers,

sensors, and other sources to derive actionable insights.

We view these framings as complementary, each emphasizing a different side of the skills and tenets required of a responsible Data Scientist. To reconcile them, we view data science as a tightly knit collection of math and computing competencies, each influencing and enhancing the other:

- **Asking questions** – Formulating questions is itself a skill, and this competency involves understanding when a question will lead to the collection of variable data, which questions can be explored with a given dataset, and what kind of data is needed to answer a question.
- **Collecting data** – An understanding that decisions are made when collecting data, and how those decisions can influence the data collected and the tools used to collect it. This competency includes knowledge of sampling, and statistical and computational mechanisms to reduce and detect variability and bias in data.
- **Visualizing data** – This includes familiarity with different data displays (e.g., pie charts, histograms, scatter plots, etc.), when it is appropriate to use them, how they can be misused, and how to interpret them. It contains both statistical and computational content, as both are involved in the construction and interpretation of these displays.
- **Analyzing data** – The ability to analyze data relies on an understanding of the variability and distribution within that data. This competency includes many statistical concepts dealing with variability and distribution, as well computational concepts for transforming (filtering, cleaning, modifying, aggregating, etc.) data.
- **Interpreting results** – Interpreting the results of an analysis requires a statistical foundation (e.g., understanding that the results obtained from a sample may vary across other samples) but also the language to accurately communicate those results to others.
- **Ethical implications** – Data science involves *making choices* about how a question is asked, how a sample is collected, what thresholds to use when grouping or analyzing the data, and the implications of building models. It considers the ways in which social biases can influence analysis, or how communicating the results of that analysis will influence society.

Each competency takes time to master, as does seeing the connection between them. In keeping with the framing of competencies found in the GAISE-II report, we think of data science class as a mix of these competencies, with varying levels of depth depending on the age of the students and the learning outcomes of the class.

### Data Literacy

The term "Data Literacy" has been used for some time, with a variety of definitions. However, in recent years more consensus has been formed around a definition that focuses on helping students become critical *consumers* of data and the results of data analysis, rather than conducting the analyses themselves. A data-literate student is aware of the data that can be collected about them and how it can be used. They are comfortable reading visualizations (charts and graphs) and summary statistics (measures of center and spread) and explaining what they convey about a dataset. In keeping with the "literacy" part of the name, we will use the term "Data Literacy" to refer primarily to the reading and interpretation of data, as opposed to the *analysis* of that data. To be clear, Data Literacy stands on its own as a valuable set of skills. One can easily imagine how being data *literate* is important when exploring the population charts or graphs in a history book, or when reading a scatter plot drawn from an experiment in science class.

Conveniently, this definition of "Data Literacy" fits comfortably into what is already happening in conventional classrooms. Most state math standards already include data visualization in middle school (e.g., reading pie and bar charts, histograms, box plots, etc.), and many schools include some material making students aware of how the data about them can be collected and used. It is no surprise, then, that many in K-12 (and even K-16!) are keen to adopt Data Literacy! However, we recognize that Data Literacy is *complementary* to the analysis that goes on in data science. Interpreting a dataset is every bit as important as knowing how to transform, filter, and otherwise interrogate that dataset – and the analysis has its own set of skills and concepts that deserve equal consideration from educators.

### Building Bridges

The slicing of Data Literacy and Data Science into "interpretation" and "analysis" domains offers a convenient lens in which to view many of the proposed standards and curricula, the question of whether data science is a math and statistics class (as framed in GAISE-II) or a computing class (as framed by industry) remains unanswered. California has rolled out new math standards, which embrace the growing support for data science as a *mathematical* concern. This has met with significant pushback and political fallout, both from the computing sphere and from the long-burning embers of the Math Wars. Providers of computing curriculum are rolling out "data science" materials, staking a claim to data science as a *computing* concern. Even at the beginning of the push for data science education, there appears to be a budding turf war between math and CS: who is the rightful "owner" of data science?

At the beginning of the chapter, we proposed a balanced approach to blending math and CS that focuses on the *conceptual bridges between the two*. And while this answer is right at home in a King Solomon allegory, we believe a broader response is necessary: **Math and computing are necessary components of responsible data science, but they are wholly insufficient**.

For example, one of the present-day harms of irresponsible data science is racially biased algorithms, which use data models trained on sentencing data that was already known to be historically biased. Instead of eliminating human bias through some notion of computational objectivity, they serve only to codify that bias. Now imagine a great Math and CS data science class, which teaches students how to ensure a representative sample of a population, and which computational methods to use when analyzing that sample. Where in this class will students consider the ethical implications of taking this sample from a population that is already filled with bias? Where will this class discuss the implications of compiling a dataset so massive that it can reliably identify the sexual orientation of individuals on social media?

Data science education must include more than just math and computing. The chapters in this book will address what "more" means in different ways and from different perspectives, but we hope that at least one takeaway is clear: we must build bridges not only between mathematics and CS, but also to other disciplines that are traditionally ignored by STEM reformers. Data science must legitimize and elevate the critical questions raised in humanities classes – not be the next movement to chip away at them.

### Data Science for Equity

This chapter began by citing two justifications for including data science in education: access to economic opportunity, and participation in democracy. Each justification presents an opportunity to enhance equity, and a risk of undermining it: will the citizens and voters of tomorrow represent the country as a whole, or merely a biased, privileged sample? Computing, data science, and data literacy are essential skill sets for many professions, especially those in science, technology, engineering, and mathematics (STEM). By 2028, the U.S. Bureau of Labor Statistics predicts that three of four new STEM job openings and three of five STEM job openings overall will be in computing. However, even when CS education is offered in K-12 schools, its results, in terms of both interest and attainment, are typically inequitable across demographic groups. As a result, students' conceptions of who can engage with the CS and data science tools necessary to solve real-world problems are insufficient for the postsecondary and professional contexts into which they will ideally move. Illustrating the consequences of differential access to and meaningful engagement with CS, the gender composition and diversity of the current CS workforce are limited, with women currently making up 25% (down from 32% in 1990) and members of underrepresented racial/ethnic minorities representing just 16% of that sector. To achieve the promise of equitable career opportunities for all, the concepts embedded in data science must have multiple points of entry, woven within classes that reach every student.

While career and degree attainment are important, we conceptualize equitable access more broadly. We move beyond mere participation in classes and careers. Even universally accessible data science will lead to inequity, if its content focuses only on statistics and computing. This book is aimed at broadening the conversation around equity. British economist Ronald H. Coarse quipped "if you torture data long enough it will confess to anything". With ethical considerations central to the above operationalization of data science, this volume explores the ways in which data science can be conceptualized as a tool for equity, diversity, justice, and inclusion. With both mathematics and computing education being necessary but insufficient ingredients of meaningful data analysis, we reflect on the purposes of data science. Beyond torturing data to say what you want it to, what ought to be the roles, goals, and responsibilities of data scientists? Within the educational context, we propose that data science can serve a transformative and central role in elucidating and illuminating historic and current spaces of inequity and injustice. Teachers are often warned that their role is to educate, not pontificate. However, challenges arise when curricular topics skirt politicized topics and those which can be spun. For example, a recent bill before the Oklahoma legislature advocated that teachers could not teach about slavery in a way that suggested that one race oppressed another through slavery or that one race was oppressed through slavery – a purportedly neutral telling of slavery wherein there were no enslaved people nor slave owners. These efforts to obfuscate history speak to the urgent need for the use of data and data science strategies for the equitable and accurate engagement of historical topics and others toward a more transparent engagement with equity. For example, typically across the United States, fourth grade students learn their state's history. However, these histories are often devoid of transparent and accurate history. For example, in California fourth grade classrooms, students learn about the Gold Rush and California Missions. These lessons often focus on the westward expansion of the United States. What is absent from these lessons is any examination of the impacts of western expansion and manifest destiny on Indigenous peoples and their lands.

Equitable data and computer science focuses on the power of identity development and agency. In moving beyond participation, this volume asks the question: In what ways can we make data and computer science education more equitable through skill development, empowerment, and the reconceptualization of who does data science. Within the chapters shared here, we see a variety of approaches to creating more accessible and equitably focused data science spaces and learning opportunities.

## References

Bargagliotti, A., Franklin, C., Arnold, P., Gould, R., Johnson, S., Perez, L., & Spangler, D. (2020). *Pre-K-12 guidelines for assessment and instruction in statistics education (GAISE) report II*. American Statistical Association and National Council of Teachers of Mathematics.

Glassdoor. (n.d.). *50 Best Jobs in America for 2022*. Glassdoor.com. https://www.glassdoor.com/List/Best-Jobs-in-America-LST_KQ0,20.htm

Miller, S., & Hughes, D. (2017). The quant crunch: How the demand for data science skills is disrupting the job market. *Burning Glass Technologies*. https://www.bhef.com/publications/quant-crunch-how-demand-data-science-skills-disrupting-job-market

IBM. (n.d.). *What is data science?* https://www.ibm.com/topics/data-science

Loews, L. (2016, September 7). *Just 1 in 4 High School Seniors Have Taken Statistics*. EdWeek. https://www.edweek.org/leadership/just-1-in-4-high-school-seniors-have-taken-statistics/2016/09

National Research Council. (2013). Next generation science standards: For states, by states.

Oracle. (n.d.). *What is data science?* https://www.oracle.com/what-is-data-science/

YouCubed. (n.d.). *Big ideas*. https://www.youcubed.org/data-big-ideas/

# 2

# PERSPECTIVES ON RESEARCH AND PRACTICE IN AND AROUND CULTURAL RELEVANCE FOR PRE-COLLEGE DATA SCIENCE IN COMPUTING

*Justice T. Walker, Amanda Barany, Alan Barrera, Michael A. Johnson, and Sayed Mohsin Reza*

## Pre-College Computing Education: Definitions and Histories

### Contemporary Computing Education

Since 1970s, when scholars such as Cynthia Solomon, Alan Kay, and others (Kay, 1977; 2011; Papert et al., 1971) were imagining the possibilities that personal computers would afford for pre-college teaching and learning, computing education has had an undeniable impact—on the tools and techniques we use to design and examine learning processes, as well as the environments within which they occur. This has spanned considerations of cognitive, social, and increasingly cultural processes. Foundational in these efforts was Seymour Papert's treatise on constructionism, and the rich opportunities that computers offer for children to expansively show and share their learning (Papert, 2020). This is reflected in research and practice efforts, ranging widely from digital activities that involve computer programs with outputs such as animations, websites, and games, to physical activities that allow learners to create complex and often shareable material artifacts to reflect their personal interests and knowledge (Kafai and Resnick, 1996). For these reasons, technical innovations in *accessible* computing have for decades provided a vast landscape for the study and support of learning in contemporary society, where digital technologies are both ubiquitous and inherent to its function.

Much of the existing research on computer science education has centered on computational thinking—a collection of practices considered to be sufficiently generalizable and usable across academic disciplines—especially in science, technology, engineering, arts, and mathematics (STEAM) (Ogegbo and

DOI: 10.4324/9781003364634-2

Ramnarain, 2022). While this framework has been important in operationalizing our understanding of learning outcomes, there have been growing efforts to broaden what we value and account for when studying what there is to gain in these areas. This has included considerations of learner self-concept, or self-appraisals of competency, ability, and potential for growth, as well as socioemotional factors such as learners' feelings and beliefs about field efficacy, agency, and belonging (Murphy and Thomas, 2008; Pellas, 2014). These measures tend to offer nuance, so that learning outcomes include not only what learners gain in terms of knowledge, but also how learning experiences may shape their long-term willingness to participate and persist. An emphasis on socioemotional features also reflects an increasingly important stance in the literature: that learning outcomes span beyond Piagtian perspectives of cognition (Piaget, 1936) toward a more Vygotskian conceptualization of learning that considers the impacts of context and environment (Vygotsky, 1978). This is reflected in the broad expansion of work in computing education that emphasizes collaboration and other social learning arrangements set in formal settings such as schools and informal spaces such as museums, camps, and makerspaces. Recognizing the apparent and inextricable link between society, culture, and cognition, scholars have advocated expanding beyond learning as computational thinking to also include a set of literacies whereby learners can read, write, and troubleshoot code, but can also interrogate the various sociocultural implications and ramifications that emerge when learners create computational artifacts (Kafai, Proctor, and Lui, 2020). This Freirian (2020) perspective brings with it a set of new priorities for designers, practitioners, and researchers in education: understanding the extent to which learners can grapple with equity tensions at the intersection of diverse social, cultural, and sociopolitical landscapes.

As digital tools and environments become increasingly adaptive, pervasive, ubiquitous in the lives of learners, this too presents both a challenge and an opportunity for culturally relevant pedagogy around computing. On the one hand, online environments have potentially revolutionized how, when, and where learning happens in the 21st century, with opportunities for customized and specialized learning around interest-driven topics that are collectively negotiated by active communities of participants (Collins and Halverson, 2018; Gee, 2017). Such participatory cultures, coupled with the modularity of digital artifacts, position learners as both producers and consumers of information (Jenkins, 2006). Despite the affordances of such technologies for supporting a "learning by doing" approach, this vision has been inconsistently and incompletely realized in the field of data science and computing education. Examinations of data science and computing education point out that most existing pedagogy offers pre-curated datasets to learners, which both obscures the motivations and intentions of the data collector, and divorces the data from students' own sociocultural contexts (Rubin, 2020). The result is inconsistent learner experience at best, and potentially damaging inequity at worst.

Research along this trajectory has raised quite unequivocally that despite extensive efforts to put computers in the hands of all children, participation in computing education has been rife with persistent inequities (Vogel, Santo, and Ching, 2017). This is reflected across learning contexts, age, and computing genres, and has spurred a number of initiatives to bring attention to these disparities and generate focus on ways to not only advance the scholarship on learning, but also advance knowledge on how we might do so equitably. Within this line of research in the United States, culturally responsive and culturally relevant pedagogies have stood out as viable lenses through which to design and study computer science based education—in part because these approaches are consistent with learning priorities in the field, and also because such an approach centers learners and their experiences as a central asset to constructing relevant knowledge (Gay, 2018; Ladson-Billings, 2008). While these approaches, like in other STEM fields, have gained significant attention in recent decades, there continues to be comparatively fewer efforts along this line in data science— marking an important watershed moment in the field wherein projects go beyond projects issues of access and empowerment to include, as a forethought, a more holistic approach to prioritize learners within the very fabric of scholarship and practice. In the next section, we characterize what we mean by data science and how the trajectory of this field has converged with efforts in computing.

### Data Science in Education

Starting with the conceptualization of the first computer-like searchable data repositories by Vannevar Bush (1945) in the 1940s, scholars have recognized that the collective sums of knowledge accumulated by and for learners was continuing to expand beyond what could be easily accessed, used, or understood. The rapid proliferation of digital tools and media in recent decades has both solved and exacerbated this problem; digital information on individuals' contexts, choices, and processes is more readily accessible than ever before, but now at scales and levels of complexity that overwhelm our traditional methods of understanding them. It is in this gap that data science emerged—an interdisciplinary field of processes, tools, and systems intended to extract insights and understanding of a phenomenon from complex, large-scale datasets that can translate into actionable models and predictions (Berman et al., 2018). If enacted intentionally, emergent data science techniques also have potential for elucidating these understandings in ways that are socioculturally situated in the skills, knowledge, identities, values, and epistemologies inherent to specific communities of practice (Shaffer et al., 2009), particularly when such work is enacted with participatory or community-based approaches (Shum et al., 2021).

An important next step in data science teaching and practice is to expand the development of these skills to K-12 learners, who will inevitably need to know how to work with data in multidisciplinary contexts across their lives and careers. While preparations for students to use and understand "big data" have not been ignored in formal and informal education, such applications have primarily focused on the specific mathematical and analytical techniques that learners might need (Lee and Wilkerson, 2018). While valuable, this leaves students ill-prepared to enact the nuanced and contextualized decision making or "thinking" that data scientists use to understand phenomena. As a result, researchers have advocated for a more humanistic approach to data science education that integrates intentional reflection on and connection to (1) the personal interests, knowledge, identities, and values a learner might bring to their work, (2) the cultural context in which the data was created, and (3) the sociopolitical considerations of ethics and power dynamics around the access, creation, and use of the data (Lee, Wilkerson, and Lanouette, 2021). It is in this way that data science education can embrace culturally relevant pedagogy as a form of learning and activism, supporting consciousness-raising, dialogue, and collaboration for and with marginalized learners (Freire, 2020; 2000).

In the next section, we offer a brief conceptualization of what we mean by culturally responsive and relevant pedagogies, as well as how we understand this perspective in relation to contemporary pre-college computer science education. Following, we engage closely with a set of illustrative vignettes of on-screen, physical, and data-based efforts in order to highlight how these perspectives have been taken up successfully to enact change and disrupt persistent field disparities—often beyond projects toward inclusion and empowerment. We end by highlighting opportunities in future research and practice that takes seriously culturally responsive and relevant pedagogies as both a learning design principle and research frame in burgeoning areas of pre-college computing—such as data science.

### Cultural Relevance and Computing Education

Culturally relevant (Ladson-Billings, 1995) and responsive (Gay, 2018) pedagogies—sometimes collectively described as culturally relevant education, or CRE (Aronson and Laughter, 2016)—were largely developed through asset-driven research methods that considered learning outcomes that were afforded when instructional designs and practices emphasized the humanization of learners, their experiences, and values. By doing so, scholars like Gloria Ladson-Billings (1995) argued that learners would be centered in learning experiences and outcomes could be exceptional in the ways she observed in cases of humanities teachers in the United States. This required educators set high expectations, and experiences include opportunities to engage in activities that are critical and of personal, cultural, or sociopolitical relevance. Consistent with this theory, Geneva Gay (2002) advocated for equipping educators with cultural competencies that are aligned with their students through continuous reflection in relation to their own power and positionality. This would enable teachers to design learning experiences that access the histories learners bring to the classroom, and also leverage those experiences as assets in the environment—important starting points and conduits on which to bridge learning with learners. Thus CRE provides resolute vantage points, and also shifts in scholarship that primarily focused on learner deficits to one that emphasizes learners as central and critical agents of learning. CRE has thus been taken up across a number of academic disciplines and, more recently, in contemporary STEM education—in order to better understand and disrupt outcome barriers that emerge after access has been established and which often disproportionately affect underrepresented and often marginalized groups such as girls and people of color (Brown, 2021; Emdin, Adjapong, and Levy, 2016; Larkin, 2019; Magee et al., 2020).

This work has been enacted along a number of fronts including scholarship aimed at understanding instructional designs (Burgess and Mensah, 2022), teacher practice (Mensah, 2009), and learning contexts such that they harness assets that are culturally familiar to learners and their lived experiences—to promote learning experiences that are quintessentially inclusive and congruent with learner identities. Examples along this include frameworks like hop-hop pedagogies (Emdin, 2013) where learning experiences occur through culturally grounded vernaculars or artistic repertoires that are common among youth (e.g., hip hop pedagogies in settings where hip hop culture flourishes). Scholarship along this frame has helped the field understand how cultural funds of knowledge and practice (Mensah, 2022) might serve a necessary entry point for learners in academic STEM disciplines that contribute to misrepresentations of science as a field of objective and rigidly systematic processes and practice. This work critically established that cultural relevance spans beyond reductive additions to learning activities (e.g., highlighting a STEM poster child or celebrating cultural day or month once a year) to encompass the rich collection of practices, identities, and communities that shape the way learners engage and understand the world. Consistent with these perspectives are other lines of CRE scholarship in STEM education and research that focuses on individual learners and the various ways we might assess success—broadening deeply cognitive measures of knowledge and practice (e.g., such as whether a learner can recite dogmatic academic language, design an experiment to answer a research question, or troubleshoot through experimental errors) to also include constructs that connect to social and cultural aspects of learning (e.g., such as the various ways learner inquiries attend to their epistemological styles, enhance their sense of self, or, relationship with community) (Brown, 2021; Larkin, 2019). Examples of work along this trajectory include examination of learner self-concept (e.g., belonging, identity, efficacy, agency, etc.) in relation to pressures (e.g., stereotype threat, microaggressions, etc.) that emerge when learning experiences are delivered independent of complex geopolitical landscapes within which learner culture and interests are shaped (Brown et al., 2020; Burgess and Mensah, 2022; Mensah, 2009; O'Leary et al., 2020). In addition, there has been significant effort placed in work that has spanned learners and learning environments to include the role family, mentors, technology, out-of-school time experiences can be understood and enacted through CRE. This scholarship has broadened and extended how the field conceptualizations and supports learners in STEM education and beyond.

Computing education is one area where CRE has had burgeoning influence on shaping scholarship discourse and encouraging activity designs that, consistent with social constructivist and constructionist traditions (Morales-Chicas et al., 2019; Frederick, Donnor, and Hatley, 2009), emphasizing prior experiences and community as a starting place for learning. These approaches have been collectively conceptualized as culturally responsive computing (CRC) and framed as approaches that center indigenous ways of knowing, diverse learner vernaculars, civic engagement, and a culture of hacking (i.e., access democratization, Eglash et al., 2013). Examples include areas like ethnocomputing (Tedre et al., 2006)—an agent-based perspective that calls for reconsideration of the cultural artifacts used when designing interfaces that otherwise narrowly reinscribe semiotic traditions and practices drawn from dominant cultures. Along this trajectory of work, scholars have brought to light the extent to which implicit design choices are embedded in social and cultural landscapes. Other instantiations of CRC include the design activities wherein learners may productively assert critique on social constructions and the myriad ways they can place groups at risk of oppression or marginalization. This approach extends ethnocomputing by situating learning in activist orientations that empower learners as change agents of cultural, linguistic, and practice traditions held by dominant, and often oppressive, cultures. Morales-Chicas et al. (2019) observed that CRC-grounded initiatives also give learners a space within which to heighten their sociopolitical awareness, express their heritage and build community. This approach frames computing as a tool that can be hacked or democratized and used to deconstruct and interrogate the world around—and toward full community or civic engagement (Eglash et al., 2013). While participation and outcome inequities persist in computing education, scholarship and practice that have taken up CRC (often as a heuristic, see Scott, Sheridan, and Clark, 2015) has shown to be critical in recasting computing scholarship—beyond projects that put computer science in the hands of all, but rather learning experiences that situate learners in a more salient part of learning design. Others have achieved important success in computing education consistent with CRC by including in their inquiries an important recognition that learners often occupy a multiplicity of intersecting identities (Akbar et al., 2022; Pinkard et al., 2017; Scott, Sheridan, and Clark, 2015) and contexts (Barron et al., 2006; Eglash et al., 2013).

Data science-based computer science education is one area where CRC work has made progress, but is inherently limited due to technical and logistical constraints. This progress is reflected, for example, in efforts to reconcile the heterogeneity of data sources and forms of educational engagement (Lee and Wilkerson, 2018). There has also been obvious recognition that learning environment designs should include opportunities for learners to engage with inquiries that are complex and connected to lived experiences. There have been a number of documented research and practice successes along this front (Lee et al., 2021), but each has helped to hone in on one inherent truth—that the contexts used to frame learning are often limited in scope and relevance since sources are restricted by data collection tools or curated by others. This seems inconsistent with and potentially antithetical to CRC because curating an inquiry is only as powerful insomuch as it can simulate real contexts and the experiences learners have each day. In the next section, we highlight illustrative case examples of where CRC perspectives have shown promise—with the goal of pointing to next directions for data science-based computing education aimed at supporting equitably productive learning. Our goal—building on Frederick et al. (2009)—is to highlight how we might design data science-based computing experiences that, consistent with CRC, enable: realistic and diverse representations, agentic student expression, authentic cultural competencies across vernaculars, and sandbox spaces for flexible engagement.

## Illustrative Case Examples

### Block-Based Programming: A Case for Technical Fluency and Creative Enterprise toward Diverse and Agentic Student Expression

Since its formal launch in 2007, Scratch has helped shape how we understand computing power and its possibilities for teaching and learning (Maloney et al., 2004). Part of the potential of Scratch as a learning tool lies in the enactment of coding in ways that Seymour, Cynthia, and Alan had imagined decades before with tools like Logo (Solomon et al., 2020)—as a space for children to learn and create. At first glance, the platform presents as a playful space for 4 to 16 year olds to explore (Maloney et al., 2010; Zhang and Nouri, 2019), with bright colors and the sprite mascot Scratch Cat. Closer inspection reveals an interface that is commonly grouped into three parts: a library or palette of coding blocks, a programming space for block assembly, and a stage where users, sometimes referred to as Scratchers, can view their creations (see Figure 2.1).

In the landscape of programming tools for education, design features such as these tend to be commonplace, and for good reason; the intuitive display

FIGURE 2.1    Scratch has a characteristic interface that is designed to support ease of entry for novice programmers. The left panel is a library or palette of coding blocks, the center panel is a programming space for block assembly, and the right panel is a stage where users can view productions.

and feature arrangement make coding accessible. Furthermore, the immediacy afforded by the creation-viewing stage means young users can practice programming, learn from their mistakes, and generate a wide number and variety of digitally dynamic artifacts. From animations, music, and art, to games, stories, and tutorials, the possibilities are expansive. Features like these have made Scratch the most frequently used platform of its kind (Zhang and Nouri, 2019) as well as a viable entry point for supporting learner technical fluencies that include computational thinking (Brennan and Resnick, 2012; Wing, 2006) and the variety of concepts, practices, and perspectives inherent to mastery. Research that has examined Scratch's utility in supporting computational thinking (broadly defined) among pre-college (i.e., K-12) groups has generally found that the tool provides ease of entry for eventual participation in a number of complex computing processes (Fagerlund et al., 2021; Zhang and Nouri, 2019). Some have attributed these affordances to design features that enable these processes: (1) low floors, or being accessible to novice coders, (2) high ceilings, or supporting students in the enactment of sophisticated activities, (3) wide walls, or encouraging a variety of styles of forms of engagement, and (4) open windows, or facilitating artifact-sharing. The emergent recognition that computer programs are socioculturally contextualized has similarly raised awareness of how the design tools and activities enabled in Scratch have deep roots and rich ties to the user communities (Walker et al., 2022). This important consideration has permeated STEM education for decades, but has only recently drawn attention in computing as scholarship increasingly highlights the ways computer programs and objects can connect to social issues (e.g., avatars that reinscribe gender or racial stereotypes, etc.) that create or exacerbate marginalization of minoritized groups.

Research that has examined these concerns highlights a spectrum of issues that threaten parity, often through limiting or restricting full participation in Scratch, ranging from predesigned one-size-fits-all commercial curriculum to underdeveloped learning activities that require prohibitive educational and technical expertise. This has led to critical work to develop learning activities that overcome these challenges and create more community-grounded learning with and within Scratch. We highlight an illustrative case example, Scratch Encore (Franklin et al., 2020), to highlight how traditional approaches to computer science learning (i.e., through on-screen activity) can be harnessed as a constructionist activity and then bridged with CRC in service of equity among learner groups traditionally underrepresented in computing. Scratch Encore (see Figure 2.2) is a set of instructional resources delivered in 15 modules; the tool is intended to serve as a curricular reframing of "traditional" Scratch applications. Titled "Multicultural,

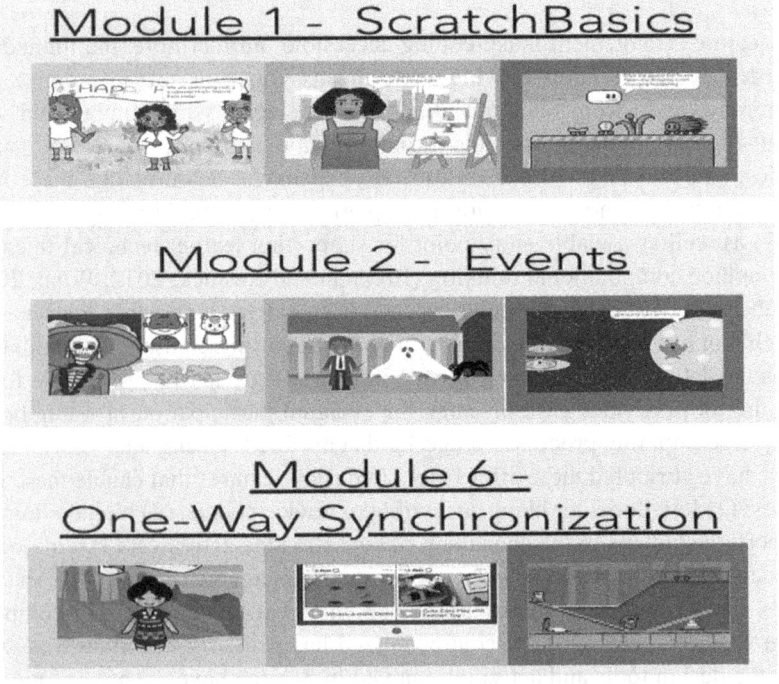

FIGURE 2.2   Scratch encore consists of modules that shift toward increased computational complexity (from top to bottom) and include opportunities for learners to use, modify, and create products that are aligned with learner cultural background, personal interests, and sociopolitical histories.

Youth Culture, and Games" on the resource site, Scratch Encore directs users to a scaffolded set of activities that embeds computing topics (e.g., conditional loops, sequence decomposition, etc.) within contextualized frames intended to encourage belonging, youth culture, and historic practice. Learners can engage with animations that are embedded in cultural traditions (e.g., the Mexican Dia de los Muertos festival) and civil rights historical events (e.g., the Million Women March). Activity modules are organized into three categories (Use, Modify, and Create) designed to strategically decrease learning scaffolds over time and support learner exploration and creative ideation around three genres (multicultural topics, youth culture topics, and games).

Franklin et al. (2020) identify three design principles that guide activities: (1) cultural relevance, (2) skill variation and exceptionalities, and (3) teacher variation. These principles are targeted to address critiques of traditional Scratch curricula in terms of promoting belonging, fluency, and teacher expertise. *Cultural relevance* is enacted through activities designed to encourage interactive relationship building with learners and teachers (e.g., through jokes) and through the inclusion of pre-designed sprites with varying skin colors and gender choices. *Skill development* is supported through learning activities such as worksheets that engage learners in the cognitive processes inquiry-driven reflections (e.g., asking questions and making predictions about code function). Activities also feature scaffolds to help learners overcome expertise thresholds; these are reflected in places where learners can use, modify, and remix existing code, or extend their inquiry and practice around a topic of personal interest. This progression from structured to more open-ended engagement provides learners with supportive, just-in-time opportunities to practice and grow toward exceptional outcomes. Importantly, Scratch Encore includes *support for teachers* that spans onboarding to curricular resources and provides access points for adapting materials in diverse learning environments. A research implementation of the program with eight teachers and nearly 300 elementary and middle-school students in the United States revealed that instructional designs supported prolonged engagement, technical fluency, and expression, reflected in the extent to which learners generated personalized objects.

This case provides insights into how on-screen computing activities can be adapted or deployed in ways that support both sophisticated learning exercises and creative expression that is tightly connected to learner culture, expertise, and personal interests. Such an approach is not only conducive to (1) computer science learning, but also encourages (2) feelings of belonging, since activity designs are couched in familiar contexts and to an extent reflect learners themselves, and (3) learner agency through the inclusion of flexible entry points that develop technical mastery and enable learners to shift from a consumer of scaffolded, pre-existing resources to a producer of their own culturally and personally relevant content. We believe this approach highlights an important piece in contemporary CRC—the notion that learner agency is as important as technical expertise and should therefore be considered a design principle in computing education.

### Physical Computing as Craft: A Case for Cultural Competencies across Vernaculars and Identities

The foundations of fabric artifacts that incorporate electronics or computer hardware (Hughes-Riley, Dias, and Cork, 2018; Kafai, Fields, and Searle, 2014) emerged with the development of tools to support their integration. Examples of these technologies include the LilyPad Arduino (Buechley and Eisenberg, 2008) that, together with conductive thread, made it possible to sew or stitch together wearable and personalizable electronics. Over the last decades this area, now commonly known as electronic or *e-textiles*, has grown from a speculative design space dominated by prototype applications of the 90s to now include a robust and flourishing community of creators who harness the power and potential of craft and computing to construct and produce expansive visions of fashion, art, and tools. This expansion of possibilities has also made significant inroads in computing education—as scholarship has attended to the myriad ways this new genre of crafting broadens what is "acceptable" knowledge and practice for computing education (Peppler, 2016). The result has been a flourishing corpus of literature that has demonstrated the potential for this craft-based genre of physical computing to support sophisticated and highly technical practices (Litts at al., 2017) across diverse modalities (e.g., on-screen programming, physical circuitry, crafting, etc.).

One successful example of this form of physical computing for pre-college learners is *Stitching the Loop* (Kafai et al., 2014), a Exploring Computer Science curricular unit that teaches learners how to design, construct, and program circuits (see Figure 2.3). The activities become progressively more complex as learners first learn circuitry by building a paper circuit powered by a coin battery and connected by copper tape. Following, they have opportunities to sew an arm bracelet using conductive thread to connect a battery holder and LEDs on a fabric arranged creatively by learners themselves. Thirdly, is a collaborative mural activity where learners use fabric or canvases to stitch single panel designs that incorporate the programmable LilyPad Arduino alongside conductive thread and LEDs. This activity is often done as a collaborative task because learners can bring together their designs and products to make one large artifact (e.g., a mural that can be displayed). Some sophisticated versions of this activity make use of light sensors so that the LED display patterns change in the presence of ambient light. In the final activity learners build a human sensor—this activity builds on the third by including conductive fabric that can be used as a sensor because it changes thread conductivity when touched. The result is a wearable

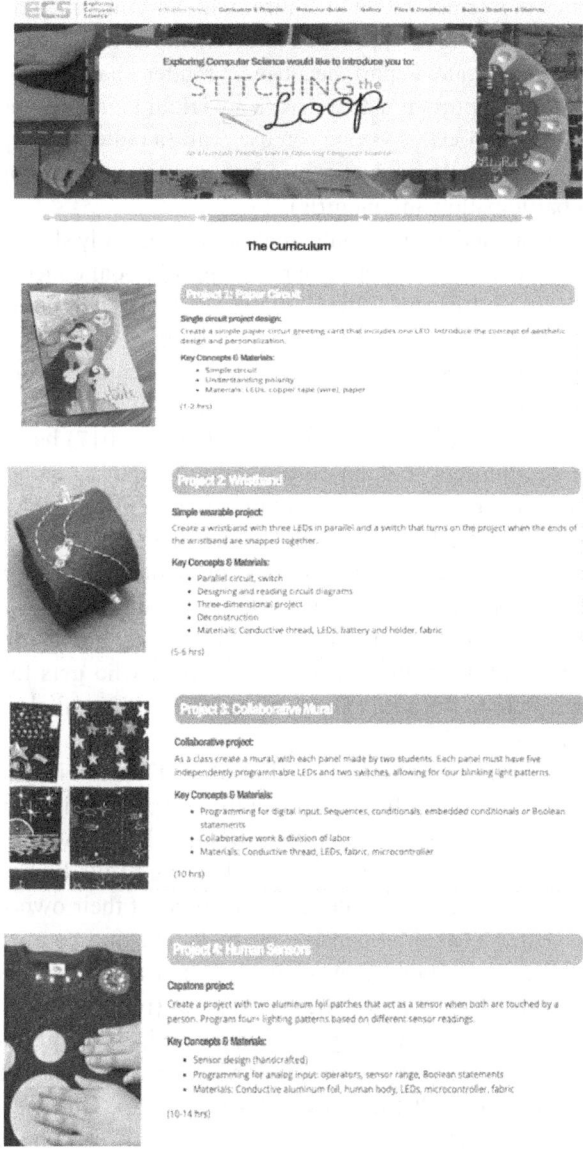

**FIGURE 2.3** Stitching the Loop is an exploring computer science e-textiles curricular unit that progressively increases in computational complexity (from top to Bottom) and engages learners with designing and constructing circuits. Projects include making a paper circuit, a wristband, a wearable garment, and an artistic mural—all of which are interactive and enabled using a battery, conductive tape, or thread, LEDS and, sometimes, a programmable Arduino.

device that users can interact with and control with varying touch patterns. This form of physical computing has prevailed as an exemplary instantiation of computing activities. This is due, in part, because it broadens participation possibilities for activity that supports practice across social and cultural vernaculars or styles, to include the various ways one can use craft (a cultural competency that has traditionally been gendered, but whose impact here has been upended as a way to support participation among girls).

Research to understand learning outcomes has consistently shown strong evidence that this approach to computing supports multimodal concepts and practices (Litts et al., 2017; Lui et al., 2016; 2020a; 2020b) in computing, circuitry, and crafting. Additionally, work has shown that e-textiles-based computer science provides a compelling frame within which to support equity efforts to broaden participation. In this regard, Kafai et al. (2019) and others (see Searle, Tofel-Grehl, and Breitenstein, 2019; Tofel-Grehl et al., 2017) have shown that learners' interests, mindsets, and appraisals about computing are sustained using this genre of computer science education—making it a viable entry point for more sophisticated academic pursuits. While these efforts mark important inroads toward achieving equity in physical computing versions of computer science education, there has been recent work to extend this scholarship toward conceptualizations that center culturally relevant computing as a primary framework that means to recast dominant narratives about who gets to participate, belong, and contribute to the field (Shaw et al., 2021a; Shaw, Fields, and Kafai, 2019; Shaw and Kafai, 2020).

Shaw, Ji et al. (2021) and Shaw, Kafai et al. (2021), for instance, take up stances consistent with culturally relevant computing perspectives in identity by framing elements of Stitching the Loop activities in relation to learner identities and perceived stereotypes about the field. In this line of work, research learners have opportunities to use e-textiles as tools to enact their own style or version of craft to represent their expansive interests, existing competencies, and knowledge to tell a story about themselves as important (existing and future) contributors to the field. In building these narratives (referred to by the authors as *restorying*), learners are encouraged to acknowledge and question relevant sociopolitical histories rife with racism. Following, learners construct their own self-narratives (Shaw et al., 2019; Shaw and Kafai, 2020). This is enacted using portfolios (shown in Figure 2.4) that simultaneously showcase their versions of history, personal futures and computing mastery (Shaw et al., 2019).

### Applied Data Science and Computing: Toward Sandbox Spaces for Flexible Engagement and Civic Priorities

Applied data science in computing can be characterized as a field that leverages computing techniques to carry out functions that enable the curation,

FIGURE 2.4   Physical computing enactments consistent with CRC perspectives on centering learner engagement in diverse vernaculars and learner identities. Shaw et al.'s (2019) report on an e-textiles intervention that created space for learners to create artifacts that recast or "[restoried] dominant CS narratives … to develop their political identities".

interrogation, and sense-making of data sets. Recognizing the steady emergence and growing popularity of this field in education contexts, scholars have sought to understand the field from diverse vantage points—including ways that emphasize the nature of data sets, practices involved with engaging them or the various contexts they inform. One prevailing instantiation of this (see Rosenberg and Jones's (2022) adaptations of Pickering, 2010) has reconciled work around data science education and research is through perspectives on agency in material, personal and disciplinary domains. *Material agency* can be described as the emergent character and nature of data as it is generated, sometimes spanning beyond what was initially conceived when it was collected. *Personal agency* refers to the emergent epistemological and, from our perspective, heuristic styles that humans exert when interacting with data sources; in education, this can be understood as how learners harness and understand data in relation to their personal experiences. *Disciplinary agency* alludes to the shared set of practices inherent to an academic discipline (e.g., mathematics, computer science, etc.) and how those norms exert influence and shape how data science is enacted.

Much of nascent data science education, especially efforts that occur in computing, is grounded in the idea that arithmetic and computational methods *in isolation* can be used to interrogate datasets in service of better understanding the world. This is a paradoxical conceptualization in many ways, because while many of the curricular frameworks that currently exist in the literature and practice do offer insights into the complex ways data science can support sophisticated and productive learning, most existing curricular applications employ restrictive, predetermined datasets that are meant to simulate, or at best replicate, real-world phenomenon. This illustrates a tension between data science-based computing and CRC, as these restrictive curricular designs are inconsistent with the need for flexible engagement that enables civic participation around key sociopolitical landscapes that impact learners and their communities. In this section, we consider a canonical example of data science curriculum (Bootstrap: Data Science, 2023) and highlight the designed potential for this type of learning. We conclude with suggestions for where the field might further develop learning structures that spur engagement that is as relevant as it is productive.

Bootstrap: Data Science (see Figure 2.5) is an illustrative example of what we mean by educational efforts that show significant promise in supporting

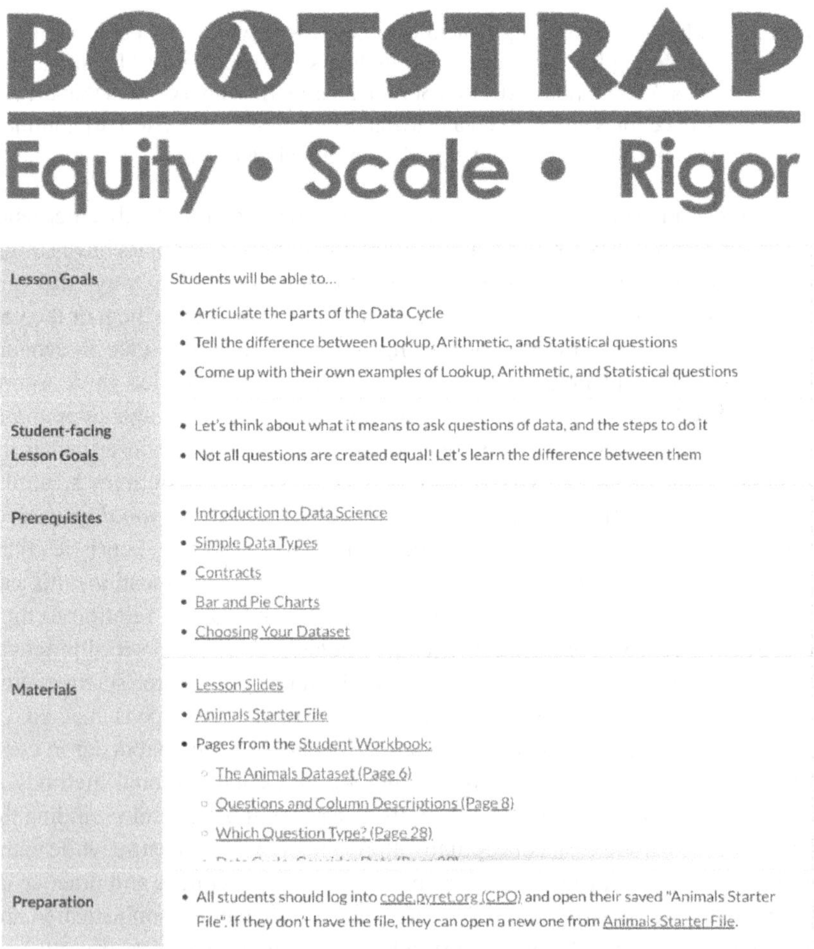

| Lesson Goals | Students will be able to... |
| --- | --- |
| | • Articulate the parts of the Data Cycle |
| | • Tell the difference between Lookup, Arithmetic, and Statistical questions |
| | • Come up with their own examples of Lookup, Arithmetic, and Statistical questions |
| Student-facing Lesson Goals | • Let's think about what it means to ask questions of data, and the steps to do it |
| | • Not all questions are created equal! Let's learn the difference between them |
| Prerequisites | • Introduction to Data Science |
| | • Simple Data Types |
| | • Contracts |
| | • Bar and Pie Charts |
| | • Choosing Your Dataset |
| Materials | • Lesson Slides |
| | • Animals Starter File |
| | • Pages from the Student Workbook: |
| |    ○ The Animals Dataset (Page 6) |
| |    ○ Questions and Column Descriptions (Page 8) |
| |    ○ Which Question Type? (Page 28) |
| Preparation | • All students should log into code.pyret.org (CPO) and open their saved "Animals Starter File". If they don't have the file, they can open a new one from Animals Starter File. |

FIGURE 2.5    Bootstrap data science offers a resource guide with curated data sources, instructional PowerPoint slides, student workbooks, worksheets and other resources to engage with computing-based data science.

productive engagement and learning from a lens of cultural relevance. As of 2019, the program has trained more than 100 teachers (Krishnamurthi et al., 2020), who applied the program in classrooms across the United States. The structure of Bootstrap: Data Science coalesces around Krishnamurthi and colleagues' (2020) three design goals: equity, scale, and rigor. Attention to *equity* is reflected in the accessibility of resources for a wide range of pre-college learners, and the tool's capacity for flexible implementation in a variety of formal and informal learning environments. The program also offers "a small amount of customization" within activities, intended to give learners a sense of ownership over their choices and actions during the experience. *Rigor* is enacted to address what creators suggest is typically missing in equity-driven programming work. This is reflected in program topics that are designed to be sufficiently difficult as to encourage struggle that is scaffolded to encourage productive outcomes. Finally, the program uses tools and teacher engagement strategies to overcome issues of *scale* that are interrupted by pesky firewalls to reach thresholds of students, including those at the margins of access. The curricular program is delivered in a traditional form with resources that include math, science and engineering standard alignment, PowerPoint presentations, student work books, and starter files with code set in a python-based coding environment. The curriculum is designed around statistics and programming spread across nine instructional units that emphases data and programming. Moreover, data sets used are "cleansed" in order to reduce programming complexity the authors characterize as prohibitive or "well beyond middle- and most high school courses" (Krishnamurthi et al., 2020).

Bootstrap: Data Science is a notable example of the designed potential of existing computer science tools, but also of what is often missing from extant curricula: connections between educational content and learner relevance (Lee et al., 2021). Other existing frameworks often fall short in providing topics and activities that support student agency (Margolis et al., 2017; Santo et al., 2019), in part because the contexts and datasets are curated by others instead of by learners for themselves. As demonstrated in this chapter, research across computing education suggests learning is most successful when situated in contexts that are familiar, relevant, and connected to personal interests, culture and sociopolitical milieu (Kafai et al., 2019; Ladson-Billings, 1995; 2014). To overcome these access and engagement challenges, we developed a data science-based CS curriculum, titled "Coding Like a Data Miner". Our curriculum was informed by culturally relevant (Ladson-Billings, 2008) and responsive (Gay, 2018) pedagogies as asset-based perspectives that constructively centers diverse learners in STEM (Brown et al., 2019; Enyedy and Mukhopadhyay, 2007; Johnson and Elliott, 2020) and connects to social and cultural assets. To achieve this, we enacted three primary features of design.

First, we adopted a participatory approach to curriculum design that invited educator and youth stakeholders to co-create the curriculum through iterative co-design sessions. This centered participants alongside researchers in critical action and reflection on curriculum design and content (Bang and Vossoughi, 2016). Students influenced curriculum topics (e.g., pop culture examples), difficulty level (e.g., scaled programming complexity), and pedagogical scaffolds (e.g., worksheets instructions). Teachers created guides for curriculum application and shaped professional development to promote flexible implementation in a variety of contexts (see Walker et al., in press). In this way, participants not only learn through the eventual sandbox experience, but also through co-design, which expands participant awareness of what sandbox possibilities exist. The result was a seventeen module learning activity (shown in Figure 2.6) that begins highly scaffolded to enhance accessibility and that eventually becomes completely free inquiry such that learners are free to pursue topics along their personal interests, cultural backgrounds, and/or sociopolitical histories.

Second, design emphasized student access to datasets from the social media platform Twitter. In line with Papert's accessibility discussions (Resnick and Silverman, 2005), Twitter provides access to data sources that are both innately familiar to social media natives (low floors), and expansive across topics and opinions, enabling users to participate along trajectories couched in their personally and culturally relevant interests (wide walls). Twitter users tend to be diverse in both opinion and context (Uddin, Imran, and Sajjad, 2014), which results in student-curated datasets that explore topics from a variety of perspectives.

Third, curriculum development and activities (exampled in Figure 2.7) were structured around the use of Twitter's application programming interface (API): a software intermediary that supports systematic access and download of large-scale datasets of tweets using search parameters learners can tailor to their needs and interests (e.g. specific topics, types of hashtags, etc.) (high ceilings). This results in a sandbox-like environment that authentically retains material, personal and disciplinary agency (Rosenberg and Jones, 2022) as learners to go beyond their typical roles as consumers of information to actively serve as producers of knowledge on their own terms with data that is real, complex, and messy. API access also necessitates the application of programming languages, providing an embedded opportunity for practical computer science skill development.

## Future Directions for Community-Engaged Research and Practice

Our examination of three illustrative example cases of computer science implementations in on-screen, physical, and data science-based computing suggest that there have been productive inroads toward making learning experiences more than accessible and empowering. These illustrations demonstrate how

BPC-DP Coding Like a Data Miner

A Culturally Relevant Data Analytics Intervention for High School Students

Project Overview    Explore Curriculum

Culturally Relevant Computing

## Explore Curriculum

Feedback Form    Curriculum Powerpoint Theme

| # | Module Category | Module Name | Slide |
|---|---|---|---|
| 1 | | Curriculum Introduction | Explore |
| 2 | | Introduction to Data Visualization | Explore |
| 3 | | Introduction to Data Analysis | Explore |
| 4 | Introduction | Introduction to Data Gathering | Explore |
| 5 | | Introduction to Statistics | Explore |
| 6 | | Introduction to Data Pre-processing | Explore |
| 7 | | Introduction to Coding | Explore |
| 8 | | Guided Inquiry 1 - Let's Gather some data | Explore |
| 9 | | Guided Inquiry 2 - Let's Do Preprocessing | Explore |
| 10 | Guided Inquiry | Guided Inquiry 3 - Let's Analyze | Explore |
| 11 | | Guided Inquiry 4 - Let's Do Statistics | Explore |
| 12 | | Guided Inquiry 5 - Let's Visualize | Explore |
| 13 | | Scaffolded Inquiry 1 - Let's Mine Twitter Data | Explore |
| 14 | Scaffolded Inquiry | Scaffolded Inquiry 2 - Let's Analyze Twitter Data | Explore |
| 15 | | Scaffolded Inquiry 3 - Let's Visualize Twitter Data | Explore |
| 16 | | Free Inquiry 1 - Chose a Topic for mining Twitter Data | Explore |
| 17 | Free Inquiry | Free Inquiry 2 - Topic Sharing and Group Discussion | Explore |

FIGURE 2.6    Coding like a data miner consists of 17 modules and uses a learning trajectory that begins highly scaffolded (to increase accessibility) and gradually becomes less scaffolded until learners reach free inquiry where they can pursue any line of inquiry couched in a personal interest, cultural significance, or sociopolitical topic or concern.

computing education can be delivered with a more nuanced frame for equity. The first case highlights how block-based programming provides initial accessibility to sophisticated technical engagement that is made possible through an abstraction of programming syntax that enables creative expression. We also observed and highlighted the ways in which coaching learning context framing and scaffolds in culturally relevant frames that connect to learner cultural backgrounds, personal interests, and sociopolitical histories provide a strong bridge

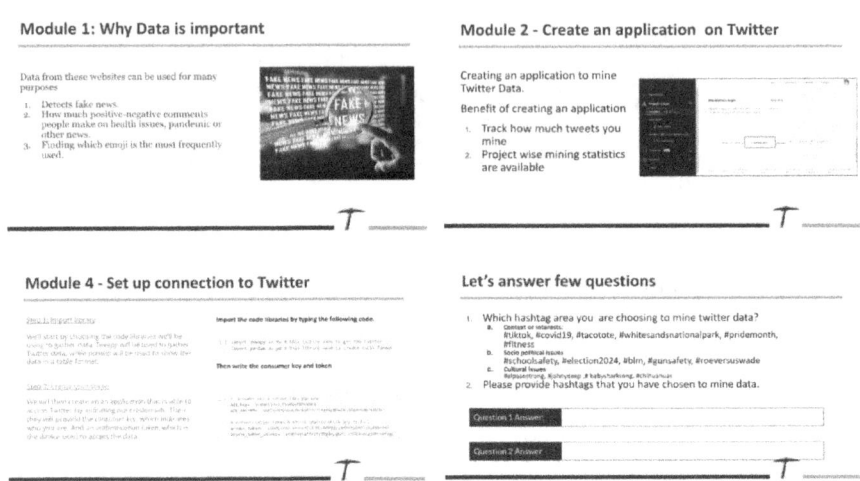

**FIGURE 2.7**   Coding like a data miner activities introduce learners to technically sophisticated coding concepts that have direct real-world application such as accessing social media-based data sets using an application programming interfaces (APIs) and conceding search queries along learner interests, cultural backgrounds, and sociopolitical histories.

toward supporting engagement and persistence after access is established. The result is a stride to supporting learner agency to express and pursue along diverse trajectories that are culturally rich, meaningful, and sustaining. Our second case highlights how physical computing as a craft broadens activities from implicit male oriented genres of engagement to include areas that are traditionally taken up by others. In doing so, craft becomes an access point for expanded practice. We also highlighted examples of instantiations of e-textiles moves beyond access projects to also include possibilities of cultural competencies that traverse diverse styles and vernaculars—and that give learners opportunities to recast narratives in ways that coincide and interact with their identities. Our third case engages with our most relevant area of computing—to both access and serve as a starting point for the vast possibilities for more equitable futures for data science. We began by highlighting a case example where material, personal, and disciplinary agency intersect to create opportunities for learners to explore data in powerful ways, and at scale. While this approach is an important step toward broadening participation to contextualized learning at myriad scales, we highlight opportunities to expand the nature of participation. We offer an example for a more sandboxed approach to data science where learners can pursue activities that are flexible and even inextricably linked to their community. These approaches highlight three frames of equity that might be considered as viable for moving beyond access and empowerment projects. We argue for data science equity perspectives that are authentic and consistent with traditional instantiations of CRC to include efforts that enable: (1) creative enterprise toward diverse and agentic student expression, (2) cultural competencies across identities and vernaculars, and (3) sandbox learning to support flexible and civic engagement. In the next section, we revisit CRC to make explicit how our articulation of equity in data science-based computing fits within and potentially expands these theoretical lenses.

CRE has and continues to be a necessary frame within which to understand equity in teaching and learning. We engage with this theory by drawing on perspectives in culturally relevant computing—CRC—to understand successes in traditional forms of computing—and where there might be opportunities to expand. We began with traditions that center learners and their diverse ways of knowing, vernaculars, civic engagement, and hacking since these starting points highlight areas of learning that span beyond access and empowerment and that emphasize learning processes. Like Morales-Chicas et al. (2019) and Frederick et al. (2009), we hold the stance that learner experiences are consequential in enacting outcomes that honor learners as change agents with cultural, linguistic, and practice talents that are typically underappreciated or underrepresented in computing. We build on work along this trajectory by examining efforts to bring data science to pre-college groups. In this examination, we highlight the ways new technologies have created opportunities to expand possibilities in data science. In light of work carried out to situate learners as data gatherers rather than reproducers in data science—we argue for learning designs that create sandbox spaces for learners to explore, curate, and flexibly engage on their own terms. Such an approach is consistent with equity emphases on learning processes and also reflects the state of the field in data science computing—where learning technologies and social media platforms allow the public to access and draw on a trove of data that spans innumerable topics and areas of exploration. The case example that uses Twitter is an illustrative account of what this means for data science learning.

## References

Akbar, M., Mortimer, K., Navarrete, G., Galvan, S., Molina, G., Reyes, R., Ontiveros, C., Gray, S., Escandon, S., Lyons, M., Delgado, P., Medrano, V., Kneedler, H., Benitez, P., Ramirez, J., Vazquez, J., & Anderson, M. (2022). The sol y agua RPP. In *SIGCSE 2022: Proceedings of the 53rd ACM technical symposium on computer science education V. 2*. (pp. 1096). https://doi.org/10.1145/3478432.3499050

Aronson, B., & Laughter, J. (2016). The theory and practice of culturally relevant education: A synthesis of research across content areas. *Review of Educational Research*, *86*(1), 163–206.

Bang, M., & Vossoughi, S. (2016). Participatory design research and educational justice: Studying learning and relations within social change making. *Cognition and Instruction*, *34*(3), 173–193.

Barron, B., Kafai, Y. B., Joseph, D., Pinkard, N., Resnick, M., Martin, C., Schatz, C., Shapiro, B., Millner, A., Peppler, K., Chiu, G., & Desai, S. (2006). Clubs, homes, and online communities as contexts for engaging youth in technological fluency building activities. In Barab, S. A., Hay, K. E., & Hickey, D. T. (Eds.), *The international conference of the learning sciences: Indiana University 2006. Proceedings of ICLS 2006, Vol. 2* (pp. 1022–1028). International Society of the Learning Sciences.

Berman, F., Rutenbar, R., Hailpern, B., Christensen, H., Davidson, S., Estrin, D., Franklin, M., Martonosi, M., Raghavan, P., Stodden, V., & Szalay, A. S. (2018). Realizing the potential of data science. *Communications of the ACM, 61*(4), 67–72.

*Bootstrap: Data Science.* https://www.bootstrapworld.org/materials/data-science/. (Retrieved 2023)

Brennan, K., & Resnick, M. (2012). New frameworks for studying and assessing the development of computational thinking. In *Proceedings of the 2012 annual meeting of the American educational research association, Vol. 1* (pp. 25).

Brown, B. A. (2021). *Science in the city: Culturally relevant STEM education.* Harvard Education Press.

Brown, B. A., Boda, P., Lemmi, C., & Monroe, X. (2019). Moving culturally relevant pedagogy from theory to practice: Exploring teachers' application of culturally relevant education in science and mathematics. *Urban Education, 54*(6), 775–803.

Brown, B., Pérez, G., Ribay, K., Boda, P. A., & Wilsey, M. (2020). Teaching culturally relevant science in virtual reality: "When a problem comes, you can solve it with science". *Journal of Science Teacher Education, 32*(1), 7–38. https://doi.org/10.1080/1046560x.2020.1778248

Buechley, L., & Eisenberg, M. (2008). The LilyPad arduino: Toward wearable engineering for everyone. *IEEE Pervasive Computing, 7*(2), 12–15.

Burgess, T., & Mensah, F. M. (2022). "He probably won't go to college": Using storytelling to illustrate how black boys use their science knowledge to challenge deficit-based teacher dispositions. In *Young, gifted and missing, Vol. 25* (pp. 37–51). Emerald Publishing Limited.

Bush, V. (1945). As we may think. *The Atlantic Monthly, 176*(1), 101–108.

Collins, A., & Halverson, R. (2018). *Rethinking education in the age of technology: The digital revolution and schooling in America.* Teachers College Press.

Eglash, R., Gilbert, J. E., & Foster, E. (2013). Toward culturally responsive computing education. *Communications of the ACM, 56*(7), 33–36.

Eglash, R., Gilbert, J. E., Taylor, V., & Geier, S. R. (2013). Culturally responsive computing in urban, after-school contexts. *Urban Education, 48*(5), 629–656. https://doi.org//10.1177/0042085913499211

Emdin, C. (2013). Pursuing the pedagogical potential of the pillars of hip-hop through urban science education. *The International Journal of Critical Pedagogy, 4*(3).

Emdin, C., Adjapong, E., & Levy, I. (2016). Hip-hop based interventions as pedagogy/therapy in STEM: A model from urban science education. *Journal for Multicultural Education, 10*(3), 307–321.

Enyedy, N., & Mukhopadhyay, S. (2007). They don't show nothing I didn't know: Emergent tensions between culturally relevant pedagogy and mathematics pedagogy. *Journal of the Learning Sciences, 2*, 139–174. https://doi.org/10.1080/10508400701193671

Fagerlund, J., Häkkinen, P., Vesisenaho, M., & Viiri, J. (2021). Computational thinking in programming with scratch in primary schools: A systematic review. *Computer Applications in Engineering Education, 29*(1), 12–28.

Fields, D. A., Kafai, Y. B., Morales-Navarro, L., & Walker, J. T. (2021). Debugging by design: A constructionist approach to high school students' crafting and coding of electronic textiles as failure artefacts. *British Journal of Educational Technology*, *52*(3), 1078–1092.

Franklin, D., Weintrop, D., Palmer, J., Coenraad, M., Cobian, M., Beck, K., & Crenshaw, Z. (2020). Scratch Encore: The design and pilot of a culturally-relevant intermediate Scratch curriculum. In *Proceedings of the 51st ACM technical symposium on computer science education* (pp. 794–800).

Frederick, R., Donnor, J. K., & Hatley, L. (2009). Culturally responsive applications of computer technologies in education: Examples of best practice. *Educational Technology*, 9–13.

Freire, P. (2000). *Pedagogy of freedom: Ethics, democracy, and civic courage*. Rowman & Littlefield Publishers.

Freire, P. (2020). Pedagogy of the oppressed. In *Toward a sociology of education* (pp. 374–386). Routledge.

Gay, G. (2018). *Culturally responsive teaching: Theory, research, and practice*. Teachers College Press.

Gay, G. (2002). Preparing for culturally responsive teaching. *Journal of Teacher Education*, *53*(2), 106–116.

Gee, J. P. (2017). *Teaching, learning, literacy in our high-risk high-tech world: A framework for becoming human*. Teachers College Press.

Hughes-Riley, T., Dias, T., & Cork, C. (2018). A historical review of the development of electronic textiles. *Fibers*, *6*(2), 34.

Jenkins, H. (2006). *Fans, bloggers, and gamers: Exploring participatory culture*. NYU Press.

Johnson, A., & Elliott, S. (2020). Culturally relevant pedagogy: A model to guide cultural transformation in STEM departments. *Journal of Microbiology and Biology Education*, *21*(1), 05.

Kafai, Y., Fields, D., & Searle, K. (2014). Electronic textiles as disruptive designs: Supporting and challenging maker activities in schools. *Harvard Educational Review*, *84*(4), 532–556.

Kafai, Y. B., Fields, D. A., Lui, D. A., Walker, J. T., Shaw, M. S., Jayathirtha, G., Nakajima, T., Goode, J., & Giang, M. T. (2019). Stitching the loop with electronic textiles: Promoting equity in high school students' competencies and perceptions of computer science. In *Proceedings of the 50th ACM technical symposium on computer science education* (pp. 1176–1182).

Kafai, Y. B., Proctor, C., & Lui, D. (2020). From theory bias to theory dialogue: Embracing cognitive, situated, and critical framings of computational thinking in K-12 CS education. *ACM Inroads*, *11*(1), 44–53.

Kafai, Y. B., & Resnick, M. (Eds.). (1996). *Constructionism in practice: Designing, thinking, and learning in a digital world*. Routledge.

Kay, A. C. (1977). Microelectronics and the personal computer. *Scientific American*, *237*(3), 230–244. https://doi.org//10.1038/scientificamerican0977-230

Kay, A. C. (2011). A personal computer for children of all ages. In *Proceedings of the ACM annual conference - volume 1* (ACM '72). Association for Computing Machinery. https://doi.org/10.1145/800193.1971922

Korhonen, T., & Lavonen, J. (2017). A new wave of learning in Finland: Get started with innovation! In S. Choo, D. Sawch, A. Villanueva, & R. Vinz (Eds.), *Educating for the 21st century: Perspectives, policies and practices from around the world* (pp. 447–467). Springer.

Krishnamurthi, S., Schanzer, E., Politz, J. G., Lerner, B. S., Fisler, K., & Dooman, S. (2020). Data science as a route to AI for middle-and high-school students. arXiv preprint arXiv:2005.01794.

Ladson-Billings, G. (1995). Toward a theory of culturally relevant pedagogy. *American Educational Research Journal, 32*(3), 465–491.

Ladson-Billings, G. (2008). Yes, but how do we do it?": Practicing culturally relevant pedagogy. In W. Ayers, G. Ladson-Billings, & G. Michie (Eds.), *City kids, city schools: More reports from the front row* (pp. 162–177). The New Press.

Ladson-Billings, G. (2014). Culturally relevant pedagogy 2.0: Aka the remix. *Harvard Educational Review, 84*(1), 74–84.

Larkin, D. B. (2019). *Teaching science in diverse classrooms: Real science for real students*. Routledge.

Lee, V. R., & Wilkerson, M. (2018). Data use by middle and secondary students in the digital age: A status report and future prospects. Commissioned Paper for the National Academies of Sciences, Engineering, and Medicine, Board on Science Education, Committee on Science Investigations and Engineering Design for Grades 6-12.

Lee, V. R., Wilkerson, M. H., & Lanouette, K. (2021). A call for a humanistic stance toward K-12 data science education. *Educational Researcher, 50*(9), 664–672.

Litts, B. K., Kafai, Y. B., Lui, D. A., Walker, J. T., & Widman, S. A. (2017). Stitching codeable circuits: High school students' learning about circuitry and coding with electronic textiles. *Journal of Science Education and Technology, 26*, 494–507.

Lui, D., Litts, B. K., Widman, S., Walker, J. T., & Kafai, Y. B. (2016). Collaborative Maker Activities in the Classroom: Case Studies of High School Student Pairs' Interactions in Designing Electronic Textiles. In *Proceedings of the 6th Annual Conference on Creativity and Fabrication in Education* (pp. 74–77).

Lui, D., Kafai, Y., Litts, B., Walker, J., & Widman, S. (2020a). Pair physical computing: High school students' practices and perceptions of collaborative coding and crafting with electronic textiles. *Computer Science Education, 30*(1), 72–101.

Lui, D., Walker, J. T., Hanna, S., Kafai, Y. B., Fields, D., & Jayathirtha, G. (2020b). Communicating computational concepts and practices within high school students' portfolios of making electronic textiles. *Interactive Learning Environments, 28*(3), 284–301.

Magee, P. A., Willey, C., Ceran, E., Price, J., & Cervantes, J. B. (2020). The affordances and challenges of enacting culturally relevant STEM pedagogy. In *Handbook of research on STEM education* (pp. 300–310). Routledge.

Maloney, J., Burd, L., Kafai, Y., Rusk, N., Silverman, B., & Resnick, M. (2004). Scratch: A sneak preview. In *Proceedings. Second International Conference on Creating, Connecting and Collaborating through Computing, 2004*. https://doi.org/10.1109/c5.2004.1314376

Maloney, J., Resnick, M., Rusk, N., Silverman, B., & Eastmond, E. (2010). The scratch programming language and environment. *ACM transactions on computing education, 10*(4), 1–15. https://doi.org/10.1145/1868358.1868363

Margolis, J., Estrella, R., Goode, J., Holme, J. J., & Nao, K. (2017). *Stuck in the shallow end: Education, race, and computing*. MIT Press.

Mensah, F. M. (2009). Confronting assumptions, biases, and stereotypes in preservice teachers' conceptualizations of science teaching through the use of book club. *Journal of Research in Science Teaching: The Official Journal of the National Association for Research in Science Teaching, 46*(9), 1041–1066.

Mensah, F. M. (2022). "Now, I see": Multicultural science curriculum as transformation and social action. *The Urban Review, 54*(1), 155–181.

Morales-Chicas, J., Castillo, M., Bernal, I., Ramos, P., & Guzman, B. L. (2019). Computing with relevance and purpose: A review of culturally relevant education in computing. *International Journal of Multicultural Education, 21*(1), 125–155.

Murphy, L., & Thomas, L. (2008). Dangers of a fixed mindset: Implications of self-theories research for computer science education. In *Proceedings of the 13th annual conference on Innovation and technology in computer science education* (pp. 271–275).

Ogegbo, A. A., & Ramnarain, U. (2022). A systematic review of computational thinking in science classrooms. *Studies in Science Education, 58*(2), 203–230. https://doi.org/10.1080/03057267.2021.1963580

O'Leary, E. S., Shapiro, C., Toma, S., Sayson, H. W., Levis-Fitzgerald, M., Johnson, T., & Sork, V. L. (2020). Creating inclusive classrooms by engaging stem faculty in culturally responsive teaching workshops. *International Journal of STEM Education, 7*(1). https://doi.org/10.1186/s40594-020-00230-7

Papert, S., Solomon, C., Soloway, E., & Spohrer, J. C. (1971). Twenty things to do with a computer. *Studying the Novice Programmer* (pp. 3–28). Routledge.

Papert, S. A. (2020). *Mindstorms: Children, computers, and powerful ideas*. Basic Books.

Pellas, N. (2014). The influence of computer self-efficacy, metacognitive self-regulation and self-esteem on student engagement in online learning programs: Evidence from the virtual world of second life. *Computers in Human Behavior, 35*, 157–170.

Peppler, K. (2016). A review of e-textiles in education and society. *Handbook of research on the societal impact of digital media*, 268–290.

Piaget, J. (1936). *Origins of intelligence in the child*. Routledge & Kegan Paul.

Pickering, A. (2010). *The mangle of practice: Time, agency, and science*. University of Chicago Press.

Pinkard, N., Erete, S., Martin, C. K., & McKinney de Royston, M. (2017). Digital youth divas: Exploring narrative-driven curriculum to spark middle school girls' interest in computational activities. *Journal of the Learning Sciences, 26*(3), 477–516. https://doi.org/10.1080/10508406.2017.1307199

Resnick, M., & Silverman, B. (2005). Some reflections on designing construction kits for kids. In *Proceedings of the 2005 conference on Interaction design and children* (pp. 117–122).

Rosenberg, J. M., & Jones, R. S. (2022). *A secret agent? K-12 data science learning through the lens of agency*. EdArXiv. https://edarxiv.org/eyzkv/download?format=pdf

Rubin, A. (2020). Learning to reason with data: How did we get here and what do we know? *Journal of the Learning Sciences, 29*(1), 154–164.

Santo, R., DeLyser, L. A., Ahn, J., Pellicone, A., Aguiar, J., & Wortel-London, S. (2019). Equity in the who, how and what of computer science education: K12 school district conceptualizations of equity in 'CS for all' initiatives. In *2019 research on equity and sustained participation in engineering, computing, and technology (RESPECT)* (pp. 1–8). IEEE.

Scott, K. A., Sheridan, K. M., & Clark, K. (2015). Culturally responsive computing: A theory revisited. *Learning, Media and Technology, 40*(4), 412–436.

Searle, K., Tofel-Grehl, C., & Breitenstein, J. (2019). Equitable engagement in STEM: Using e-textiles to challenge the positioning of non-dominant girls in school science. *International Journal of Multicultural Education, 21*(1), 42–61.

Shaffer, D. W., Hatfield, D., Svarovsky, G. N., Nash, P., Nulty, A., Bagley, E., Frank, K. A., Rupp, A. A., & Mislevy, R. J. (2009). Epistemic network analysis: A prototype for 21st-century assessment of learning. *International Journal of Learning and Media, 1*(2), 33–53.

Shaw, M. S., Fields, D. A., & Kafai, Y. B. (2019). Connecting with computer science: Electronic textile portfolios as ideational identity resources for high school students. *International Journal of Multicultural Education, 21*(1), 22–41.

Shaw, M., & Kafai, Y. (2020). Charting the identity turn in K-12 computer science education: Developing more inclusive learning pathways for identities. In *The Interdisciplinarity of the Learning Sciences, 14th International Conference of the Learning Sciences (ICLS) 2020*, Gresalfi, M. & Horn, I. S. (Eds.) (pp. 114–121). International Society of the Learning Sciences.

Shaw, M. S., Ji, G., Zhang, Y., & Kafai, Y. B. (2021). Promoting socio-political identification with computer science: How high school youth restory their identities through electronic textile quilts. In *2021 Conference on research in equitable and sustained participation in engineering, computing, and technology (RESPECT)* (pp. 1–8). IEEE.

Shaw, M. S., Kafai, Y. B., Zhang, Y., Ji, G., Russo, R., & Aftab, A. (2021). Connecting with computer science: Two case studies of restorying CS identity with electronic textile quilts. In *Proceedings of the 15th international conference of the learning sciences-ICLS 2021*. International Society of the Learning Sciences.

Shum, S. B., Irgens, G. A., Moots, H., Phillips, M., Shah, M., Vega, H., & Wooldridge, A. (2021). Participatory Quantitative Ethnography. In *International conference on quantitative ethnography proceedings supplement* (pp. 126–138).

Solomon, C., Harvey, B., Kahn, K., Lieberman, H., Miller, M. L., Minsky, M., Papert, A., & Silverman, B. (2020). History of logo. *Proceedings of the ACM on programming languages, 4*(HOPL), 1–66. https://doi.org/10.1145/3386329

Tedre, M., Sutinen, E., Kähkönen, E., & Kommers, P. (2006). Ethnocomputing: ICT in cultural and social context. *Communications of the ACM, 49*(1), 126–130. https://doi.org/10.1145/1107458.1107466

Tofel-Grehl, C., Fields, D., Searle, K., Maahs-Fladung, C., Feldon, D., Gu, G., & Sun, C. (2017). Electrifying engagement in middle school science class: Improving student interest through e-textiles. *Journal of Science Education and Technology, 26*, 406–417.

Uddin, M. M., Imran, M., & Sajjad, H. (2014). Understanding types of users on Twitter. arXiv preprint arXiv:1406.1335.

Vogel, S., Santo, R., & Ching, D. (2017). Visions of computer science education: Unpacking arguments for and projected impacts of CS4All initiatives. In *Proceedings of the 2017 ACM SIGCSE technical symposium on computer science education* (pp. 609–614).

Vygotsky, L. S. (1978). *Mind in society: The development of higher psychological processes*. Harvard University Press.

Walker, J. T., Stamato, L., Asgarali-Hoffman, S. N., Hamidi, F., & Scheifele, L. Z. (2022). Community laboratories in the United States: BioMakerspaces for life science learning. *Public Understanding of Science*, 09636625221135858.

Wing, J. M. (2006). Computational thinking. *Communications of the ACM, 49*(3), 33–35.

Zhang, L. C., & Nouri, J. (2019). A systematic review of learning computational thinking through scratch in k-9. *Computers & Education, 141*, 103607. https://doi.org/10.1016/j.compedu.2019.103607

# 3

# SHRINKING LANDS AND GROWING PERSPECTIVES

## Affordances of Data Science Literacy During a Culturally Responsive Maker Project

*Tyler Hansen, Kristin Searle, Mengying Jiang, and Melissa Barker*

## Introduction

As our society becomes increasingly saturated with data and individuals leave digital traces of themselves, often unknowingly and unintentionally, educators must consider how students engage with and learn about data. As it is currently construed, data science tends to "center white, masculine, and Western norms and values" (Lee Wilkerson, and Lanouette, 2021, p. 1098), resulting in biased data (Eubanks, 2018; Noble, 2018). Because data is often biased, questions about how students learn about data in relation to issues of race, class, gender, and power are particularly important (e.g., D'Ignazio and Klein, 2020; Philip, Olivares-Pasillas, and Rocha, 2016), yet remain relatively understudied (Lee et al., 2021).

Data science education is related to but distinct from computer science (CS) education. Data science remains an emerging, interdisciplinary field, but it is set apart from other disciplines that deal with data through its focus on "data collected in an incidental or automated manner from extensive social and environmental contexts" (Wilkerson and Polman, 2020, p. 1, citing Donoho, 2017), such as social network data or data from remote sensors. Data scientists then use the tools of CS to collect, clean, analyze, model, visualize, and otherwise create narratives with data. Importantly, there is a tension within data science as, on the one hand, a discipline that is designed to advance workforce development and national economic interests and, on the other, a tool of democratic participation (Philip, Schuler-Brown, and Way, 2013). Thus, data science education must not only attend to the necessary mathematical, statistical, and computational tools and methods, but also when, where, how, and from whom data is collected.

DOI: 10.4324/9781003364634-3

Students must see themselves as data scientists and recognize the personal, cultural, and sociopolitical dimensions of data.

Like scholars thinking about data science education, CS education research has recently taken a more critical turn, focusing on justice-centered (e.g., Lachney et al., 2021; Vakil, 2018) and culturally responsive sustaining approaches (Kapor Center, 2021) to computing education that begin from a place of critical reflection on systems of oppression and our own places within those systems and then leverage that awareness to engage students in understanding how computing is connected to their own identities, cultures, and communities, and can be used as a tool for taking sociopolitical action. In this way, CS education is connected to data science education, but focused on a more specific set of disciplinary skills, including thinking like a computer and programming, that facilitate aspects of data science.

In this chapter, we present findings from a pilot classroom implementation of the *Shrinking Lands Project*, which seeks to help middle school students visualize and understand the loss of Indigenous lands in the United States as a result of colonization. This project was designed as part of a larger curricular design effort focused on bringing together Montana's Indian Education For All standards, which require all K-12 students in Montana to learn about 12 Tribes of Montana, with CS standards. The *Shrinking Lands Project* integrates data science, computing, and making to engage students in a hands-on visualization and exploration of how tribal lands in Montana decreased in area over time through government to government treaties and land theft. Using electronic textiles (e-textiles) materials, middle school students created data visualizations to show how Blackfeet Nation land was reduced through treaties with the U.S. government beginning in 1851, with sewn, programmable circuits showing how much land was lost with each subsequent treaty. The data visualization begins with 10 LED lights turned on across four felt land pieces and ends with no LED lights turned on because current Blackfeet reservation lands are less than 10 percent of the Blackfeet Nation's ancestral homelands. In this chapter, we examine student level data about how middle schoolers made sense of both personal data and distal data (data pertaining to social, cultural, or political topics), noting potential changes in understanding throughout this curricular unit. We engaged in a discourse analysis of student questioning to explore the potential affordances of the making and computing for scaffolding and supporting student understanding of data and data science.

## Literature Review

As the call for K-12 data science education has grown, so too have efforts to create meaningful frameworks for thinking about data science education (e.g., Lee et al., 2021; 2022). In this chapter, we bring together the humanistic stance toward data science education put forth by Lee et al. (2021) with the Kapor Center's framework for culturally responsive sustaining computer science education (2021). Because our larger curriculum design project is focused on CS education, we begin by explicating the importance of culturally responsive sustaining CS education and then articulate how we view this approach connecting with a humanistic stance toward data science education.

Culturally responsive sustaining computing grows out of work in culturally relevant pedagogy (Ladson-Billings, 1995), culturally responsive teaching (Gay, 2018), and ethnomathematics (Tedre et al., 2006). Initially, culturally responsive computing focused on providing relevant STEM learning experiences, supporting learners in their cultural identities, and lessening the distance between STEM disciplines and learners' experiences (Eglash et al. 2013) through the design of computational tools and learning environments. Building on this work, recent efforts have focused more on pedagogy that supports students in using computing as a tool for sociopolitical critique and sociopolitical action. For instance, Scott, Sheridan, and Clark (2015) suggest that we must attend to students' intersectional identities as we "consider who creates, for whom, and to what ends rather than who endures socially and culturally irrelevant curriculum" (p. 421). Even more recently, the Kapor Center defines culturally responsive sustaining CS pedagogy as pedagogy that:

> ensures that students' interests, identities, and cultures are embraced and validated, students develop knowledge of computing content and its utility in the world, strong CS identities are developed, and students engage in larger socio-political critiques about technology's purpose, potential, and impact.
>
> *(2021, p. 5)*

In the emphasis on connecting to students' interests, identities, and cultures and in the emphasis on developing students' awareness of sociopolitical contexts and their abilities to critically assess the work technology does in the world, we see overlap with taking a humanistic stance toward data science education.

Lee et al. (2021) argue that as educational researchers increasingly think about how to engage students in learning about data science, we need to think through what engagements with data should look like across three layers which interact with practices of collecting, analyzing, and visualizing data because "any educational data activity is in fact a product of manifold individual and social tools and processes" (p. 665). At the personal layer, we are concerned with students' direct involvement in data practices and their reasoning about data. While engagement with personal data may lead to a greater sense of agency and ownership of data, students may also be more concerned with the trustworthiness of data given their proximity to data collection processes. At the cultural level, we are concerned with how the available tools, artifacts, and social norms of a given context guide activities around data. As Lee et al. (2021) note, "the cultural layer can subtly but substantially shape what is measured and how, what types of patterns can be uncovered and described, and how investigators collect, calibrate, evaluate, and communicate data and findings" (p. 666). The cultural layer may also shape students' sense of what counts as data, what stories can be told about data, and what sources of data are trustworthy. Finally, the third layer is the sociopolitical layer, which is concerned with how specific data sets and data practices speak to larger discourses of power and privilege. This layer speaks most directly to calls to engage students in examining and challenging inequitable power structures in our society (Louie, 2022), but remains the most understudied of the layers.

In this pilot study we brought together culturally responsive sustaining CS education and a humanistic stance on data science education through hands-on making of data visualizations designed to show the loss of tribal lands over time. Making activities like e-textiles have shown particular promise for engaging students in academic disciplines that might be unfamiliar by leveraging familiar craft materials and making practices (e.g., Kafai et al., 2014). Through the creation of computational artifacts, youth are supported in recognizing their own capacities to generate meaning and create things. For instance, Stornaiuolo (2020) reported on high school youth generating their own datasets and then creating screen-printed visualizations of their data on t-shirts and tote bags. Students developed a sense of data ownership by seeing themselves an agentive data producers and they were able to see data as contextualized resources that have potential for multiple purposes. However, students tended to focus their projects on understanding themselves in new ways rather than on issues of sociopolitical importance, demonstrating the challenges of engaging students in sociopolitical critique in both CS and data science education. Therefore, this study explores the ways in which students make sense of data at both a personal and sociopolitical layer, and the ways in which they bridge the familiarity gap within the context of data visualizations, historical injustices, and CS.

## Frameworks

Aspects of this curriculum were designed through various related frameworks. The CS aspects of the curriculum were guided by the Kapor Center's Framework for Culturally Responsive Sustaining Computer Science Education (Kapor Center, 2021) while the framing understanding of data science engaged Lee et al.'s (2021) humanistic understanding of data science. The Montana Office of Public Instruction's framework for implementing the Essential Understandings Regarding Montana Indians (MOPI, n.d.) guided the curricular design of indigenous content in conjunction with the voices of indigenous community members.

Within the Kapor Center Framework, there are six key components. Educators must be both aware of their own racial positionality and be willing to actively dismantle racism within CS education. While this pilot was conducted outside of Montana, similarities between Montana Tribes' loss of land over time and the experiences of Tribes within the state that the pilot took place in were discussed in class. Educators must also create a culture of equity and inclusion within their classroom that foster CS opportunities. The personal data aspect of the curriculum provided opportunities for students to explore their own interests, identities, and cultures within the context of data science. Next, both curriculum and pedagogical strategies must not only align with standards, but also with student interest and background by using CS to critically examine the sociopolitical implications in which CS is embedded. In addition to the personal data aspects of the curriculum, students were guided both directly via instruction on how data is used to form a particular narrative that is created by those that generate and interpret data, and indirectly by visualization of the loss of tribal lands over time. Fourth, students are given their own voice and agency within the classroom. The final product of the curriculum provided an opportunity to create their own data visualizations on topics of their choosing. Further, opportunities were provided for students to explore other types of data within the context of Montana Tribes' loss of land over time. Fifth, family and community assets are intentionally incorporated into the curriculum. Part of exploring personal data was asking students to think about how they use data outside of school and making projects that included data collected from out-of-school experiences. Finally, a variety of voices are incorporated into the classroom for students to hear from a diverse background about the potential ways in which they could pursue CS as a career. The nature of personal data afforded students, teachers, and researchers to discuss future career pathways in CS and data analysis.

In addition to the Kapor Center's framework on CS, this project was created by using the three layers of humanistic data work (Lee et al., 2021), which suggest that students should engage with data at personal, cultural, and sociopolitical levels to make experiences with data both meaningful and equitable. Personal data refers to a student's direct involvement with data. This can either refer to a student's observations or their direct involvement with data collection and interpretation. The curriculum included multiple opportunities where students could explore personal data. For example, students created their own data visualizations from data about their personal lives (e.g., number of pets). The cultural layer refers to a wide array of community data practices. It is often interpreted as the tools used by data scientists; however, we must also consider how we collectively gather, interpret, and communicate information. By not only exploring personal data, but finding creative ways to convey data allowed students to think about the ways in which we collectively interpret and convey information. For example, the personal data collected by students prompted a discussion on the difference between correlation and causation. The sociopolitical layer contends that the ways we use data is ultimately determined by power structures. From mass surveillance to white supremacy, examples of this are many. In the case of this curriculum, students explore tribal lands shrinking over time. This provides a data-driven counter-narrative to hegemonic retellings of U.S. history.

Here, we evaluate our implementation by using these frameworks as a guide. This approach allows us to identify discrepancies between the curriculum and framework, as well as gain deeper insights into the practical applications of these frameworks. Ultimately, this enables us to develop strategies for future implementations. With these frameworks as our guiding principles, we asked the following research questions:

1  Did students improve their understanding of personal and distal data over time?
2  What are the discursive affordances of making when discussing equity within a data science context?

## Methods

### Context

#### Montana as a Context: Indian Education and Computing for All

The *Shrinking Lands Project* is part of a larger curriculum and professional development project designed to bring together Montana's Indian Education For All Standards and it recently adopted CS standards in the context of social studies. The State of Montana, in its 1972 Constitution (Article X, Section 1(2)), acknowledged the distinct cultural heritage of American Indians and made a commitment to offer education that would safeguard the cultural integrity of every Montana Tribe. In 1999, the Legislature passed House Bill 528, which established the law known as Indian Education for All. Subsequently, in 2005, the Legislature allocated the initial funding to assist Montana schools in fulfilling this long standing commitment (MOPI, n.d.). We designed an integrated social studies curriculum to engage middle school students with sociopolitical issues surrounding indigenous lands in Montana through data visualizations and computing. The state of Montana implemented their Computer Science (CS) Standards in July 2021. A crucial component of these standards is data science and visualization, with data and analysis being a key standard starting from the elementary school level. For example, grade 5 students are expected to visually organize and present data to highlight relationships and substantiate claims.

*The Shrinking Lands Project*

*The Shrinking Lands Project* is a six lesson e-textiles data visualization project that explores how the historic tribal lands of the Blackfeet Nation in Montana were impacted by treaties made with and broken by the U.S. government. E-textiles combine sewing, circuitry, and CS by having students sew LED lights, batteries, and microcontrollers together with conductive thread to create a functional, programmable circuit.

**Project Design.** The purpose of the project was to provide students with a data-driven perspective of the historic changes to land autonomy experienced by the Blackfeet Tribe through changing and dishonored treaties with the U.S. government. The quantities of Blackfeet Nation land from 1850 to present were obtained via tribal government websites and the U.S. Geological Survey. Those areas were calculated by using Google Earth's measure function. We then created maps of each treaty, proportional to the size of each land mass on felt. Examples of these felt pieces are shown in Figure 3.1. Students then used ratios to calculate the percentage lost with each treaty and assigned the appropriate

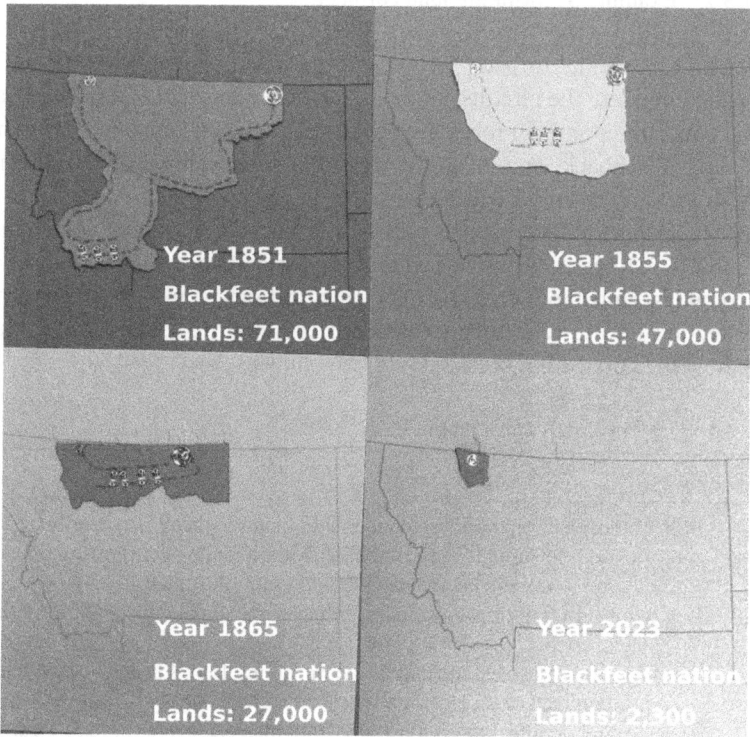

**FIGURE 3.1**   Picture of shrinking lands project.

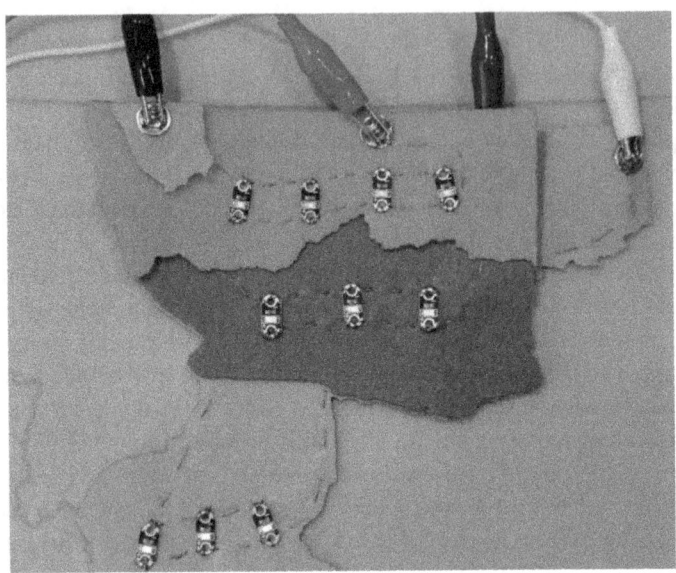

**FIGURE 3.2** Example of shrinking lands project.

number of LED lights to each felt land piece. Using snaps to unite the land pieces and complete the circuits, students were able to situate Blackfeet Nation lands at four time points in relation to the land of the current state of Montana (see Searle, Fischback, and Tofel-Grehl, 2022). An example of the Shrinking Lands project is shown in Figure 3.2.

The Shrinking Lands project is part of a larger middle school social studies unit focused on treaties and sovereignty. It consists of six one hour long lessons focused on learning about data and data visualizations, constructing the project, and reflecting on the relevance of both historic and personal data. Table 3.1 shows how the curriculum breaks down over six lessons.

**TABLE 3.1** Shrinking lands curriculum

| Day(s) | Activities |
| --- | --- |
| 1 | Project introduction, defining data, making simple data visualizations (bar charts) to show loss of Blackfeet Nation lands over time |
| 2 | Introduction to circuitry and working with e-textiles materials |
| 3 | Begin construction of projects; discussion on personal data and correlation vs. causation |
| 4–5 | Construction and programming of the Shrinking Lands project using felt, LEDs, conductive thread, and microcontrollers |
| 6 | Making connections between personal and historical data through Dear Data inspired visualizations of class data |

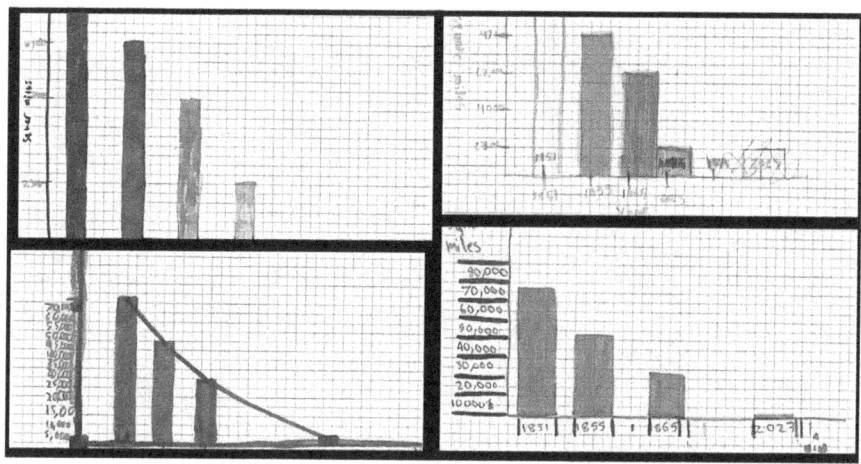

**FIGURE 3.3**    Examples of bar charts.

The first day consisted of an introduction to the project and the concepts of data and data visualization. Students used markers and graph paper to make simple data visualizations representing the land area lost by the Blackfeet Nation through a series of treaties with the U.S government (Figure 3.3). On the second day, students were introduced to the circuitry skills needed to create data visualizations with e-textiles. Throughout the first two lessons, we noticed that the historical data was very abstract for students. Because research suggests that the idea of data is more concrete when it is connected to students' personal lives (Stornaiuolo, 2020), students were also asked to think about data science as it is related to their personal lives. To explicitly illustrate a connection between personal data and ways in which historical data can be interpreted, a member of the research team used students' personal data to demonstrate the limitations of data. Specifically, the researcher illustrated this by showing that students with blonde hair had more pets than students of any other hair color. After writing the results on the board, the researcher led a discussion on the difference between correlation and causation. This activity engaged the class in a discussion about inferences and issues of sampling error (Figure 3.4). This anchored a class discussion about how inappropriate inferences can be made from the stories data tells. Students were then asked to use what they learned through the activity and discussion to generate their own questions about the Shrinking Lands project and the Blackfeet Nation that could be answered through data, thus making connections between students' personal data and distal data. On Days 4–5, students created their e-textile data visualizations with felt, LEDs, conductive thread, and microcontrollers. Each piece of felt represented the amount of Blackfeet land remaining after a given treaty and had LED lights on it corresponding to what

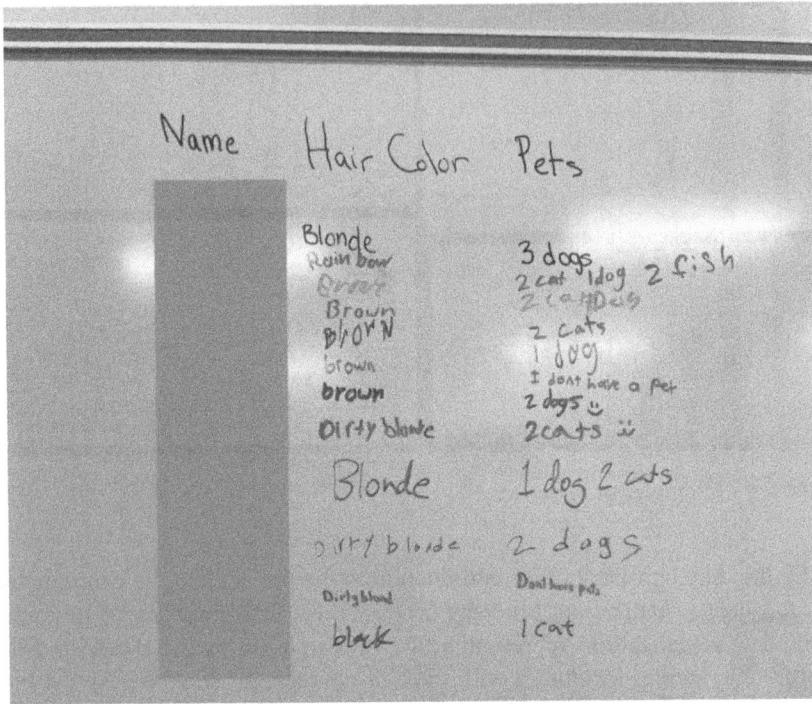

**FIGURE 3.4**    Picture of personal data students put on the whiteboard.

percentage of original tribal lands remained. When students put together their felt land pieces, they created data visualizations of the shrinking land area of the Blackfeet Nation over time. To visualize the proportion of land lost, students used Circuit Playground Express microcontrollers to program LED light bulbs to turn off with each treaty. Initially, all 10 lights on the project are lit up and then multiple lights turn off with each treaty (each light represents 10 percent of the initial area) until the present day when there are no lights because it is less than 10 percent of original Blackfeet lands. As students completed construction and programming of their felt land pieces, they were asked to explore the data visualization they created in relation to a series of guiding questions developed by the research team and online resources showing the changing landscape of tribal lands in Montana through various treaties. On Day 6, students were able to explore their own personal data by creating visualizations of their own. These data visualizations were inspired by Dear Data (http://www.dear-data.com/theproject). Dear Data is a project that began with two information designers, Giorgia Lupi and Stefanie Posavec, who would exchange personal analog data via postcards. They would gather personal data, find creative ways of conveying that data, create legends or guides to interpret the data, and ship them to each other. Students were able to look at Dear Data examples to create visualizations using class data about things like pets, hobbies, and favorite type of music.

## Setting

The project was piloted in a grade 7 classroom in a suburban charter school in the western United States. Importantly, this site was not in the state of Montana due to logistical considerations. The state in which the data was collected has a similar historical narrative of white settlement and the loss of Indigenous lands, which students were learning about in their social studies class at the time of data collection. The project took place over six consecutive Fridays where we met with students for 55 minutes each time. Of 38 students participated in the project, with 14 male students and 24 female students. The majority of students were white and none of them indigenous.

## Data Collection

### Surveys

To gain understanding about how middle school students think about data on a personal, cultural, and sociopolitical level (Lee et al., 2021), and to understand how they made sense of the Blackfeet Nation's land reduction through treaties over time, we administered a survey consisting of seven open-ended questions on Day 2 and again on Day 6. The questions were designed after we noticed how much students struggled with the idea of data in our first session to gain a better sense of how students make sense of both types of data from pre to post. The first question in the survey asked students to define data, the next three questions asked about their own personal data, and the last three questions asked about historical data and possible interpretations of data through a sociopolitical lens (e.g., how the data we have can be limited by the people that collect it). It is important to note that students did have some instruction about data in the first period before they were given the initial survey measure.

### Discourse

In addition to the surveys, we audio recorded each class period. Each table within the classroom had a recorder on it, and research team members also carried recorders to document their specific conversations with students. We used an automated transcription service for initial transcription and then a member of the research team listened to every audio file and corrected the transcription until it matched all classroom utterances as accurately as possible.

### Quantitative Analysis

We hypothesized that students would improve their understanding of both proximal and distal data over the course of the project. To answer our first research question about whether students could identify and describe personal and distal data over time, we first had to aggregate and quantify the survey questions. We removed any missing data from our sample. This left us with a sample of $n = 31$. Two coders were involved throughout the entire process and a third researcher provided guidance and context during the first phase of coding. First, we open coded the survey responses to understand the nature of student responses to the questions (Corbin and Strauss, 1990). We identified two questions that asked students about their personal data, proximal, and two questions that asked students to describe data use beyond or outside themselves within a social, cultural, or political context, distal data. The proximal questions were "How do you use data in your own life?" and "What kinds of data could you collect about yourself?". The distal questions were "How might we use data (your own or historical data) to tell stories?" and "What kinds of data do you think are missing from the historical record? Why?" (Table 3.2).

Our coding system was a simple binary (yes or no) on whether the students provided evidence of understanding what personal data is for each question in both pre and post surveys, and whether students provided evidence for understanding of what distal data is. For this initial analysis we defined evidence for understanding of personal data as describing any kind of data and explaining a personal connection to it. A student who responded with "my swimming averages" was coded as a yes since they demonstrated the ability to utilize personal data from their own life to gain new insights. On the other hand, a student who provided the answer "video games" was coded as a no because the response was too vague to discern the student's thoughts or intentions regarding any information related to the topic. While there is plenty of data that could be gathered and used in video games, the answer does not explicitly describe a form of data.

Defining and coding distal data was more difficult, as we wanted to look at how students made sense of data in cultural and justice-centered contexts. Therefore, we defined a "yes" for distal data when the student mentioned a

**TABLE 3.2** Questions used for quantitative analysis by category

| | |
|---|---|
| Questions pertaining to personal data | How do you use data in your own life? |
| | What kinds of data could you collect about yourself? |
| Questions pertaining to distal data | How might we use data (your own or historical data) to tell stories? |
| | What kinds of data do you think are missing from the historical record? Why? |

cultural, political, or societal issue, and mentioned any form of data that directly pertains to that issue. For example, "You don't always hear both sides of a story. For example, in war, the war is told from the perspective of the victor" would be coded as a yes because it both pertains to data (stories) and is within a sociopolitical context (perspective of the victor). On the other hand, the response "migrants because of history" would be coded as no. While this response does, at least vaguely, mention a societal problem, it fails to specify a type of data that pertains to it. We coded each answer to all four questions as yes or no which gave us quantifiable descriptive data for making sense of the potential growth of students. Finally, we coded student answers to determine if they were able to broadly explain personal and distal data. Since we were only looking for macro-level understandings of personal and distal data, we coded yes for any instance of describing those data types. For example, if a student was coded yes for a question pertaining to personal data but coded for no on the other personal data question, the student was coded as a yes overall. We achieved an interrater reliability of 100% by coding together, highlighting difficult answers, referring to our definitions, and debating answers until a conclusion was reached.

The frequencies for each category were tallied and analyzed. We used a chi square test for independence to determine the relationship between the understanding of both types of data from pre to post.

## Qualitative Analysis

For the second research question about the discursive affordances surrounding equity by use of this curriculum, we first employed a macro-level discourse analysis. Specifically, we analyzed student questioning, as students were encouraged to form their own questions about social issues that could be answered by data throughout the project. While we used the Kapor Center's framework on CS and Lee et al. (2021) as our guide, we also initially open coded all student questions to see if any other themes that did not pertain to frameworks were also present (Corbin and Strauss, 1990).

We documented every question students asked over the six days of instruction into an Excel file. Two reviewers read sections of questions and coded them as one of the following: N-nonsense (a question that had nothing to do with the project or data), A-activity (questions related to how they should complete the physical task they were engaged in), B-Blackfeet (a question related to the Blackfeet Tribe), D-data (a question about data in general), BD-Blackfeet Data (a question specific to data related to the Blackfeet Tribe) E-equity (a question about diversity, injustice, equity, or other sociopolitical topic, not specific to the Blackfeet Tribe) (Table 3.3).

The first code, "nonsense" related to questions students asked that were unrelated to the project, such as "Does anyone else know what Wario is?". The

**TABLE 3.3**  Codes used to describe the variety of questions posed by students during this maker project

| Code | Letter | Meaning |
|------|--------|---------|
| 1 | N | Nonsense – unrelated, irrelevant |
| 2 | A | Activity – related to maker activity |
| 3 | B | Blackfeet – question about the Montana Blackfeet Tribe |
| 4 | D | Data – general question about data |
| 5 | BD | Blackfeet Data – Montana Tribe and their data |
| 6 | E | Novel, culturally situated questioning surrounding equity, diversity, inclusion, or justice |

second code, "activity", involved questions pertaining to the making or construction aspects of the project. These questions were almost entirely procedural questions (e.g., "How do I make it light up?"), but also consisted of a small number of conceptual questions about circuitry or coding. The third code, "Blackfeet", related to general questions about Montana Tribes that could not be answered by data and did not discuss Tribes in a sociopolitical or cultural context. For example, the question "What is the difference between Blackfeet and Blackfoot?" was coded as a question about Montana Tribes. While this question does pertain to Blackfeet Nation, it does not ask a question that could be answered with data. Additionally, the question did not involve diversity, equity, inclusion, injustice, ideology, social movements, public opinion, or power and therefore could not be considered a question about sociopolitical issues. The fourth code, "data", was defined as questions about either students' personal data or questions that could be answered with data, but not within a sociopolitical or cultural context. For example, one student asked another student "How much blood are you using right now? How many words have you written? How much lead have you used while writing?". This student clearly listed questions about different types of data (in this case, personal data). While the data did not pertain to Blackfeet Nation or questions of sociopolitical topics, they still were asking questions that could be answered by collecting and interpreting data. The fifth code, "Blackfeet Data", was connected to questions about Montana Tribes that could be answered with data, or questions generated about Montana Tribes informed by data. For example, a student asked "Did [Blackfeet Nation] have enough resources?" when inquiring about the consequences of the shrinking lands. Not only could this question be answered with data (quantifying resources), but also was asked about Blackfeet Nation specifically. Finally, the sixth code, "equity", related to novel questions that were data-driven, culturally or sociopolitically situated, but beyond the scope of Montana Tribes. For example, one student asked if they could use data visualizations to convey "How many people caught COVID?". This question could be answered by data (how many), is rooted in a sociopolitical or cultural context (pandemics are a social phenomenon by nature), and is beyond the scope of what the students were taught (the *Shrinking Lands project* does not involve pandemics at all).

The reviewers met together, reviewed the coding of the other, and resolved any conflicts about how questions should be coded. Because a large part of this curriculum encouraged students to generate their own questions about Blackfeet Nation lands, similar happenings in their own state's history, and other questions surrounding justice and equity, we decided to analyze the questions within the discourse specifically. However, after our initial immersion in the data, we also typed in other forms of discourse that were relevant to the study. Specifically, we analyzed the conversations that centered on engaging with data in a culturally situated context. We did this to provide more context about student questions. For example, if a student asked "it's the same with [site's state] Tribes, right?", we would code the discourse leading up to and following that question to both better understand the nature of the question and the specific discourse that led to it.

Two coders coded all questions. 100% interrater reliability was achieved by coding together, highlighting tough calls, and resolving disagreements with our agreed upon definitions. Lastly, we tallied the frequencies of each code by day. Daily macro-level analysis gave us a broad understanding of the nature of the discourse throughout each day, allowing us to make comparisons between the types of student questioning and the nature of the curriculum that day.

## Results

### Research Question 1

A chi-square test of independence was performed to examine the differences in student understanding of proximal data and distal data from pre-assessment to post-assessment (Table 3.4). The relationship between scores from pre-assessment to post-assessment was not statistically significant, ($X^2$ (3, 62) = 4.66, $p$ = .19, Cramer's $V$ = .27) which indicates there were no significant differences in student understanding from pre-assessment to post-assessment.

While there were no significant differences from pre- to post-assessment, there are a few results from this chi-square test that are worth noting. First, adjusted standardized residuals show that there were significant shifts in student understanding of proximal data, but not understanding of distal data from pre to post ($p$ = .03). Most students already understood proximal data at the time of the pre-assessment, but did not understand distal data. Second, there were also shifts from pre- to post-assessment in student understanding of both proximal and distal data ($p$ = .06). In other words, there were statistically significant increases in student understanding of distal data in the post-assessment. While there was an

**TABLE 3.4** Frequencies and chi square test for independence

| Evidence for Data Understanding | Pre-assessment | Post-assessment | $X^2$ | $\varphi c$ |
|---|---|---|---|---|
| Yes proximal, no distal | 23 (2.08**) | 15 (−2.08**) | 4.66 | .27 |
| No proximal, yes distal | 0 | 0 | | |
| No proximal, no distal | 4 (−.69) | 6 (.69) | | |
| Yes proximal, yes distal | 4 (−1.82*) | 10 (1.82*) | | |

Note: **$p$ < .05. *$p$ < 0.1. Adjusted standardized residuals appear in parentheses next to group frequencies.

increase in the number of students who did not demonstrate an understanding of either proximal or distal data, this shift was insignificant ($p = .49$). Finally, our Cramer's V with three degrees of freedom is a medium effect ($\varphi c = 0.27$), per Cohen (1988). Given these results, our interpretation is that significant changes did happen in specific instances, but were not significant overall from pre-assessment to post-assessment. One impacting factor of statistical significance might be the very limited sample size. Another might be students' levels of participation; while most students were present at all sessions, some were absent repeatedly. These small shifts might account for lower knowledge gains within an already small sample. So while findings are not significant, we are encouraged by the noticeable shifts in understanding across some students.

### Research Question 2

Overall frequencies for each question type can be found in Table 3.5. The majority of questions that were asked pertained to the making aspect of the project. While the quality of these questions varied, it is undeniable that the majority of students were motivated by making. One might argue that asking how to complete a particular task is obviously going to be the most prevalent. However, students asked these questions to teachers, researchers, and each other unprompted, implying at least some motivation. Further, while students displayed reluctance to ask

**TABLE 3.5** Frequency of each student questioning type overall

| Question Type Code | Frequency | Percent |
|---|---|---|
| Nonsense – unrelated, irrelevant | 426 | 30 |
| Activity – related to maker activity | 858 | 60 |
| Blackfeet – question about the Montana Blackfeet Tribe | 29 | 2 |
| Data – general question about data | 15 | 1 |
| Blackfeet Data – Montana Tribe and their data | 90 | 6 |
| Novel, data-related, culturally situated questioning | 8 | <1 |

questions about the Blackfeet Nation in group discussions, even when prompted, during the making and construction portion of the class this was not the case.

One of the key findings from the coding of the classroom discourse was that students were motivated to understand data through the context of their project. While students were quiet and disengaged during several of the class discussions on the data, students engaged readily when it came to constructing computational circuit representations of the changing lands accessed by the Blackfeet Nation. This included more obvious examples of motivation, with comments like "Why are you having so much fun in school today?", but also afforded a more collaborative dynamic during construction. Students asked each other questions such as "Do you know where I can get the information so I can make it?" In relation to their e-textiles projects, students said things like, "This way is funner", and "It's more hands-on and interactive". Beyond the large frequency of activity-related questions, students overwhelmingly understood the potential benefits of more creative methods of data visualizations. This ranged from simple aesthetics, such as stating that "It has more color" or "The lights are shiny", to more nuanced observations about the nature of the project allowing students to see and interpret data more meaningfully. For example, a student noted that "It shows [the data] and you can actually see how much they lost". Another student echoed this sentiment, "It has actual shapes". We infer that comments like these indicate the motivating aspects of this curriculum, and that this motivation provides an overall better medium for students to make sense of data. One student asked "What is the [smallest piece of felt]?". This demonstrates a motivation to understand data that wasn't present while students were making bar graphs of the same information. Students never asked what the smallest bar was on the bar graphs. We can reasonably argue here that the making aspects of the curriculum gave students a hands-on means of interpreting data. Some students would only interpret the data visualization at a basic level, like "They lose like a ton of land; all of their land". However, our argument is best captured by a student reflecting on their project. While looking at the current land area for Blackfeet Nation, they said "It's so small that the light doesn't fit". Through the recognition that the light couldn't fit within the representation of the diminished land area, students are actively engaging with the profound implications of their data visualization. Their realization that the model fails to adequately depict the amount of remaining land highlights the stark reality of Blackfeet Nation. By making this project, students were able to more easily visualize data and make more meaningful inferences about the data they were given. For example, students would ask each other "where does this [light] go?" to make their projects work. They would have to have at least a basic understanding of the data that their project was visualizing.

Students were able to ask novel questions pertaining to culturally situated data, albeit infrequently. Several students asked questions about the Blackfeet Nation's resources. For instance, one student asked "did they have enough resources?". This implies thought and considerations about the implications of these various treaties and subsequent land theft. However, some students took it a step further to explore culturally situated data as it related to their own interests. Other students inquired about the specifics of the treaties, such as asking "did they have to have a passport (to leave)?". Some students took the next step by exploring hunting rights. A student asked "I thought they had a treaty that said they had enough to hunt?", followed by, "I don't know what their hunting rights are". That student spent his spare time in class trying to find out more information about the hunting rights of Blackfeet Nation. As the student had amassed this data, other students wanted to know more. Another student asked him a few days later "Where did you find the deer data?". This implies interest because students were seeking out cultural and sociopolitical data independently. Other students commented on the similarities of racial discrimination experienced between Black Americans and Indigenous peoples. We infer that the culturally situated nature of the curriculum provided opportunities to discuss broad and specific issues of race, culture, equity, and justice based on data.

Motivation also played into the generation of novel student thoughts and questioning. For example, one student asked, "Can we do it again with a different Native American Tribe from [state name]?". While this was inspired by teachers discussing the connections between different Indigenous Nations, arguably it was the student's desire to do more maker projects combined with the knowledge that other Tribes are affected in a similar way. Students making data-driven comparisons between Indigenous Nations was the most common way they thought of novel questions. While exploring Blackfeet Nation data, a student asked, "It's the same with the [Dakota], right?". However, by learning about the creative ways in which data can be visualized, students generated other types of data that could be visualized in the same way. This ranged from simple observations about the general lack of creative data displays, like "Why has no one done [more creative data visualizations]?". Other students expressed interests in similar projects on different topics, such as, "How many people caught COVID?" or investigating the lives of "Wild West culture". This demonstrates that students not only are thinking about data in their own interests, but also are actively thinking of ways that they can integrate maker data visualizations.

## Discussion and Conclusion

This study examines the piloting of the *Shrinking Lands project*. Our findings suggest a promising beginning implementation and suggest several areas where the project could be improved. While we did not find statistical significance in changes in student understanding of data overall, we found that students improved their understanding of distal data. Further, qualitative results indicate student motivation as a primary affordance of using maker projects for data visualization.

Students were able to make connections around issues of equity using e-textiles. Additionally, students were able to make meaningful data visualizations from e-textiles. The results overall lead us to argue that maker projects serve as a useful tool for students when creating data visualizations to engage discussions around historic inequity. Making also seems to be a way that students can more easily discuss data. This makes intuitive sense, as these projects are tangible.

Our quantitative results showed a change in student understanding of distal data over the course of the project. Most students already understood personal data pretty well, which is consistent with other studies (e.g., Acker and Bowler, 2018). It is interesting that none of the students were able to identify distal data if they couldn't identify proximal data. Given the motivating aspects of making and students searching for different ways to convey data with making, it stands to reason that adding other data surrounding Blackfeet Nation of the student's choosing to the existing visualization, or making a new visualization with e-textiles could prove effective for bridging the gap between understanding personal data and distal data.

Our qualitative results indicate motivation to learn the material through the maker activity. This is largely unsurprising and consistent with other research (e.g., Kafai et al., 2014). However, we were also able to demonstrate that students made sense of data through the maker activity. While many maker activities involve CS (e.g., Tofel-Grehl et al., 2017), few have incorporated making into data visualization. These results are promising in pursuing this avenue further. Given the overlap between calls for humanistic data (Lee et al., 2021) and culturally responsive computing (Kapor Center, 2021), it would make sense that making activities could apply well to both areas.

## References

Acker, A., & Bowler, L. (2018). Youth data literacy: Teen perspectives on data created with social media and mobile devices. *Proceedings of the 51st Hawaii international conference on system sciences*, *9*, 1923–1932.

Cohen, J. (1988). *Statistical power analysis for the behavioral sciences* (2nd ed.). Erlbaum.

Corbin, J. M., & Strauss, A. (1990). Grounded theory research: Procedures, canons, and evaluative criteria. *Qualitative Sociology*, *13*(1), 3–21.

D'ignazio, C., & Klein, L. F. (2020). Seven intersectional feminist principles for equitable and actionable COVID-19 data. *Big Data & Society*, *7*(2), 2053951720942544.

Eglash, R., Gilbert, J. E., Taylor, V., & Geier, S. R. (2013). Culturally responsive computing in urban, after-school contexts: Two approaches. *Urban Education*, *48*(5), 629–656.

Eubanks, V. (2018). Automating inequality: How high-tech tools profile, police, and punish the poor. St. Martin's Press.

Gay, G. (2018). Culturally responsive teaching: Theory, research, and practice. Teachers College Press.

Kafai, Y., Searle, K., Martinez, C., & Brayboy, B. (2014). Ethnocomputing with electronic textiles: Culturally responsive open design to broaden participation in computing in American Indian youth and communities. In *Proceedings of the 45th ACM technical symposium on Computer science education* (pp. 241–246).

Kapor Center. (2021). *Culturally responsive-sustaining computer science education: A framework.* https://www.kaporcenter.org/wp-content/uploads/2021/07/KC21004_ECSFramework-Re

Lachney, M., Bennett, A. G., Eglash, R., Yadav, A., & Moudgalya, S. (2021). Teaching in an open village: a case study on culturally responsive computing in compulsory education. *Computer Science Education, 31*(4), 462–488.

Ladson-Billings, G. (1995). Toward a theory of culturally relevant pedagogy. *American Educational Research Journal, 32*(3), 465–491.

Lee, V. R., Wilkerson, M. H., & Lanouette, K. (2021). A call for a humanistic stance toward K–12 data science education. *Educational Researcher, 50*(9), 664–672.

Louie, J. (2022). Critical data literacy: Creating a more just world with data. In: *Proceedings of the National Academy of Sciences' Workshop on Foundations of Data Science for Students in Grades K-12* (pp. 1–16). 13–14 September. National Academy of Sciences.

Noble, S.U. (2018). *Algorithms of oppression: How search engines reinforce racism.* New York University Press.

Philip, T. M., Schuler-Brown, S., & Way, W. (2013). A framework for learning about big data with mobile technologies for democratic participation: Possibilities, limitations, and unanticipated obstacles. *Technology, Knowledge and Learning, 18*, 103–120. https://doi.org/10.1007/s10758-013-9202-4

Philip, T. M., Olivares-Pasillas, M. C., & Rocha, J. (2016). Becoming racially literate about data and data-literate about race: Data visualizations in the classroom as a site of racial-ideological micro-contestations. *Cognition and Instruction, 34*(4), 361–388.

Scott, K. A., Sheridan, K. M., & Clark, K. (2015). Culturally responsive computing: A theory revisited. *Learning, Media and Technology, 40*(4), 412–436.

Searle, K. A., & Fischback, L. (2022). Using Electronic Textiles to Visualize the Loss of Tribal Lands Over Time. In *Proceedings of the 16th International Conference of the Learning Sciences-ICLS 2022* (pp. 1874–1875). International Society of the Learning Sciences.

Stornaiuolo, A. (2020). Authoring data stories in a media makerspace: Adolescents developing critical data literacies. *Journal of the Learning Sciences, 29*(1), 81–103.

Tedre, M., Sutinen, E., Kähkönen, E., & Kommers, P. (2006). Ethnocomputing: ICT in cultural and social context. *Communications of the ACM, 49*(1), 126–130.

Tofel-Grehl, C., Fields, D., Searle, K., Maahs-Fladung, C., Feldon, D., Gu, G., & Sun, C. (2017). Electrifying engagement in middle school science class: Improving student interest through e-textiles. *Journal of Science Education and Technology, 26*, 406–417.

Vakil, S. (2018). Ethics, identity, and political vision: Toward a justice-centered approach to equity in computer science education. *Harvard Educational Review, 88*(1), 26–52.

Wilkerson, M. H., & Polman, J. L. (2020). Situating data science: Exploring how relationships to data shape learning. *Journal of the Learning Sciences, 29*(1), 1–10.

# 4

# DESIGN OF TOOLS AND LEARNING ENVIRONMENTS FOR EQUITABLE COMPUTER SCIENCE + DATA SCIENCE EDUCATION

*Shuchi Grover, Devin Jean, Brian Broll, Veronica Cateté, Isabelle Gransbury, Akos Ledeczi, and Tiffany Barnes*

## Introduction

In response to the nationwide Computer Science for All or "CSForAll" (Smith, 2016) movement, K-12 schools and classrooms across the United States have seen tremendous growth in the teaching and learning of computer science (CS) in this last decade. This growth has been fueled by three-pronged goals—(1) a recognition of a world increasingly infused with computers and computer-related applications that run all aspects of our lives; (2) a sharpened focus on broadening participation in computing in both classrooms and industry aimed at women and students from ethnic minority groups that have historically been marginalized from CS with a view to creating a diverse and inclusive workforce; and (3) a need to promote problem solving and computational thinking (CT)/doing skills among learners as key skills to address the challenges of the 21st century. This movement has thus resulted, among other things, in large-scale training of teachers to teach CS, design of curricula for all levels of CS learning in primary and secondary schools, learner-friendly programming environments and tools, and the creation of a new advanced placement (AP) high school course—AP CS Principles (Astrachan, 2011)—designed with the specific goal of making CS inviting for all teens, promoting interest for continuing to learn CS in college, and providing a foundation for success in future CS learning.

Even as the report card on achievement of these goals is a mixed bag (Code. org, 2022; Google LLC & Gallup Inc., 2020) and the movement redoubles efforts toward increasingly equitable and culturally relevant approaches to CS education to effect much-needed improvement in diversity and inclusion in computing classrooms (e.g., Davis et al., 2021; Morales-Chicas et al., 2019),

DOI: 10.4324/9781003364634-4

there is now a growing movement to also make room for data science (DS) as a school subject and especially in CS (Krishnamurthi and Fisler, 2020; Rosenberg et al., 2020; Srikant and Aggarwal, 2017). The US National Science Foundation now recognizes a "pressing need for education and training programs that equip future generations with skills in both CT—the ability to develop computational abstractions of real-world phenomena—and inferential thinking—the ability to arrive at sound conclusions from data" (NSF, 2017). One big motivation for this push is the monumental growth of datasets that is powering an unprecedented revolution in AI and machine learning. With DS growing as a discipline in CS departments in colleges, and big data affecting critical aspects of our lives, learning DS in K-12 as part of, or in conjunction with, CS offers students opportunities to develop skills in programming, topics of statistics, critical thinking, communication, and also consider the social implications of data and their use. Given the increasing role of data in so many realms of human endeavor, from finance and commerce to healthcare, entertainment, and technology it is essential that all students—especially those from groups historically underrepresented in STEM—have access to DS in their school learning.

For all the aforementioned reasons, researchers are increasingly arguing for rethinking the content of introductory computing around a data-centric approach to better engage and support a diversity of students (Krishnamurthi and Fisler, 2020). It is worth noting that fostering the learning of CT especially in interdisciplinary subject integration settings often relies on data (Grover, 2021a, 2021b). Based on prior empirical research as well as relevant literature on CT integration projects, Grover (2021b) proposed the *CTIntegration* framework that argues for leveraging data as a linchpin or vehicle for curricular integration of computing and programming with other subjects. Such interdisciplinary experiences involve students engaging in data practices (describing, examining, processing, visualizing, interpreting) with data from (or relevant to) the domain, to interrogate phenomena, engage in sense-making, and answer questions from the domain. Some examples of such data-driven CT and CS integration include Bootstrap BS:DS (Krishnamurthi and Fisler, 2020), creating computational artifacts to visualize the great migration of Blacks from the American South during the Jim Crow era using physical computing in social science classrooms (Cannell, Tofel-Grehl, and Searle, 2020), and integration of CT into environmental systems science in middle school (Grover et al., 2020). Engagement with data in interdisciplinary learning supports diversity and inclusion in STEM learning by providing multiple entry points for engagement engendered by the variety of interdisciplinary projects across many domains (Grover, 2021b; Israel et al., 2020; Mahmoud, 2005) and also prepares learners for new topics such as AI (Srikant and Aggarwal, 2017). Such integration is all the more relevant since CS is not the only subject affected by this DS education revolution. New standards for science learning, such as the Next Generation Science Standards (NGSS), also place data analysis and interpretation as one of the core science and engineering practices (Lead States, 2013), and the use of "big data" in K-12 classrooms is on the cutting edge of science education (Kastens, Krumhansl, and Baker, 2015). Mathematics teaching and learning at the secondary level is also undergoing a tectonic shift with DS and statistics superseding advanced math topics such as algebra, trigonometry, and calculus at the high school level (Erickson et al., 2019; Finzer, 2013). Clearly, these trends prompt us to rethink the tools as well as CS curricular experiences, especially at the high school level, when students have the necessary foundational math competencies and introductory programming skills for robust engagement with data and computing.

### Learning Environments for Addressing the Need for Rich and Equitable CS+DS

A growing body of literature describes the features of tools and learning environments that support equitable learning opportunities to students from groups that have been historically marginalized in computing (Madkins et al., 2020). These include lowering the barrier to programming through accessible tools and curricula that serve the needs and interests of all students, pedagogies (such as project-based learning (PBL), guided inquiry, and collaborative activities) that make STEM learning more welcoming to women and minoritized groups, and culturally relevant projects and activities that not only help make connections to students' backgrounds, cultures assets, and communities but also engage students in critical interrogation of societal inequities and issues that impact human liberties and rights. Past research suggests that female and underrepresented students are drawn to socially relevant projects and people-oriented fields (Ceci et al., 2014; Fisher and Margolis, 2003; Kirk et al., 2012; Papastergiou, 2008). Framing the learning of such CS and DS, or "CS+DS", topics in interdisciplinary ways (through curricular integration) also allow for multiple entry points for learners (Grover, 2021b), making the experience more appealing to girls and students from minoritized groups—both of whom have been shown to enjoy projects that represent meaningful and socially consequential activity (Madkins et al., 2020; Margolis and Fisher, 2002). Such interdisciplinary learning experiences can thus also be leveraged for achieving the goals of equitable teaching and learning of computing.

Block-based programming in open-ended programming environments is arguably the most popular and accessible vehicle for introducing K-12 students to programming and CS while also attending to the crucial goals of promoting interest and creativity (Fields et al., 2016), as well as increasing diversity (Weintrop, Killen, & Franke, 2018). However, the most popular block-based programming environments in use in K-12 curricula today (such as Scratch, Snap!, and MIT App Inventor) lack a critical component of CS+DS experiences—a uniform and intuitive way to access and work with real-world data resources on the internet (Brady et al., 2022).

Against this backdrop, we present our efforts to develop and enrich tools and learning environments that organically and equitably integrate CS+DS learning, especially at the high school level. We draw examples from a collaborative, multi-institutional NSF-funded research and development project called *CS Frontiers (CSF)* that focuses on creating curricula to engage high school girls in emerging subfields at the frontiers of computing. The CSF curriculum includes curricular modules aimed at engaging high school students in exciting, data-driven inquiry projects that also help build an understanding of advanced CS topics: (1) distributed/networked computing, (2) cybersecurity and Internet of Things, (3) AI and machine learning (AI/ML), and (4) software engineering. CSF modules leverage NetsBlox (Broll et al., 2017), an extension of Snap! that provides seamless, intuitive, and easy-to-use block-based programming experiences involving integration of data and datasets, and also helps promote learning of advanced topics in CS through the use of a plethora of freely available web services and APIs. In the following sections, we present features and design principles of NetsBlox, as well as pedagogical principles from STEM and CS education research that guide our design of equitable data-centered learning experiences, followed by two curriculum examples that we have piloted: (1) an interdisciplinary experience that integrates CS and DS in service of authentic data-driven inquiry in climate science and (2) a socially relevant AI and machine learning curricular module.

## NetsBlox and Design Features Supporting Data Science, Equity, and Frontiers of CS+DS

NetsBlox, built on the open source codebase of Snap!, introduces a simple abstraction to provide access to online data sources and web services. *Remote procedure calls (RPC)* allow users to invoke functions running remotely on the NetsBlox server and provide results as return values that correspond to variable types native to the Snap! environment (e.g., numbers, strings, images, and lists). Related RPCs are grouped into *Services*. Examples include Google Maps, Weather, the Movie Database, Climate, and many others. Many of their RPCs wrap web API calls to third-party providers. RPCs use a single, self-documenting block named "call" that has two pull-down menus, one for the Service and one for the RPCs. Figure 4.1 shows some sample RPC calls.

On the back of this simple abstraction, NetsBlox opens up the world of distributed computing and facilitates inquiry into data-driven engagement with interdisciplinary ideas and topics. In this section, we detail many affordances of NetsBlox[1] that are aligned to the cause of equity in STEM classrooms and allow us to design curricula and learning environments to support inclusive STEM and

```
call  WordGuess ▼  /  guess ▼  word
call  Chart ▼  /  draw ▼  lines  options
call  MovieDB ▼  /  movieRuntime ▼  id
call  Translation ▼  /  translate ▼  text  from  to
call  Twitter ▼  /  recentTweets ▼  screenName  count
call  CloudVariables ▼  /  getVariable ▼  name  password
call  NewYorkTimes ▼  /  getMostViewedArticles ▼  period ▼
call  COVID-19 ▼  /  getConfirmedCounts ▼  country  state  city
call  IceCoreData ▼  /  getCarbonDioxideData ▼  core  startyear  endyear
```

**FIGURE 4.1**   Example remote procedure calls (RPC) calls.

CS+DS learning opportunities. To exemplify these design principles, we share code snippets from various CSF projects that have been designed and piloted with about 150 students over several summer camps in 2020–2022 facilitated by 10 high schools CS teachers who have been trained through our PD.

### Low Floor in Block-Based Programming

In an influential 2005 paper, Resnick and Silverman outlined the guiding principles for designing constructionist tools for young learners. One of the core principles emphasized the need for tools to be easy for novices to get started (low floor). Block-based programming in general, and in particular, Snap! (a Scratch derivative), have been designed for providing gentle introductory programming experiences to K-12 learners. Research (e.g., Goldenberg et al., 2020; Weintrop et al., 2019) shows that block-based programming benefits minoritized students and those with little-to-no "preparatory privilege" (Margolis et al., 2008) and prior programming knowledge.

NetsBlox follows in the low-floor tradition of Snap! Even students who have never used NetsBlox before can learn the basics of RPCs and build their first distributed app in one class period. The simple RPC block also embodies another Resnick and Silverman (2005) mantra, "a little bit of programming goes a long way". CSF modules typically introduce students to NetsBlox RPC features through simple yet powerful weather and climate projects. Figure 4.2 shows the simplicity of the code to effect amazing outcomes. Note how easy it is to plot a visualization showing information about atmospheric carbon dioxide

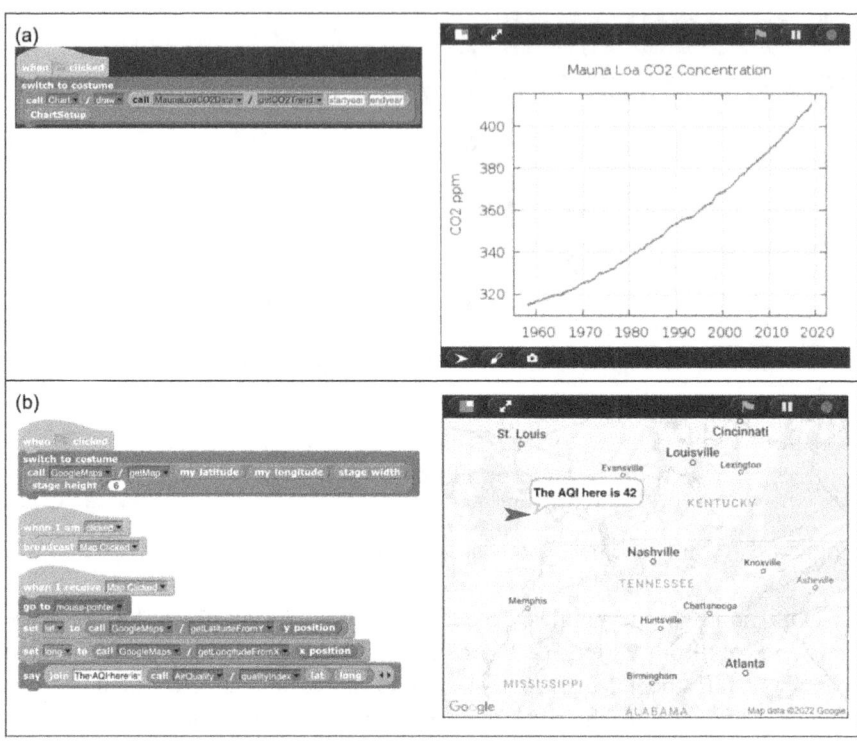

**FIGURE 4.2** Low floor of NetsBlox allows students to easily access and visualize real datasets. (a) $CO_2$ concentration visualization over time. (b) Instantaneous air quality readings on Google Map

concentrations measured on Mauna Loa in Hawaii by the NOAA (Figure 4.2a). Figure 4.2b shows the code to simply and instantaneously provide readouts of Air Quality at any location clicked on a Google map. The reactions these early experiences (achieved with such low effort) elicit from students is often visceral.

### Access to Real-World, Authentic, and Relevant Datasets

NetsBlox is constantly evolving, with new web-based services being added and updated regularly. Many of these services provide simplified interfaces to various web APIs and datasets that span several subjects and topics including (but not limited to) various science topics, astronomy, Twitter, New York Times, climate science, geolocation and maps, databases for images, movies, and lyrics, Project Gutenberg, Language APIs for text analysis and translation, AI APIs for Natural Language Processing or NLP (such as CodeNLP and Parallel Dots) and "ThisXDoesNotExist" (generative AI at https://thisxdoesnotexist.com/), Alexa,

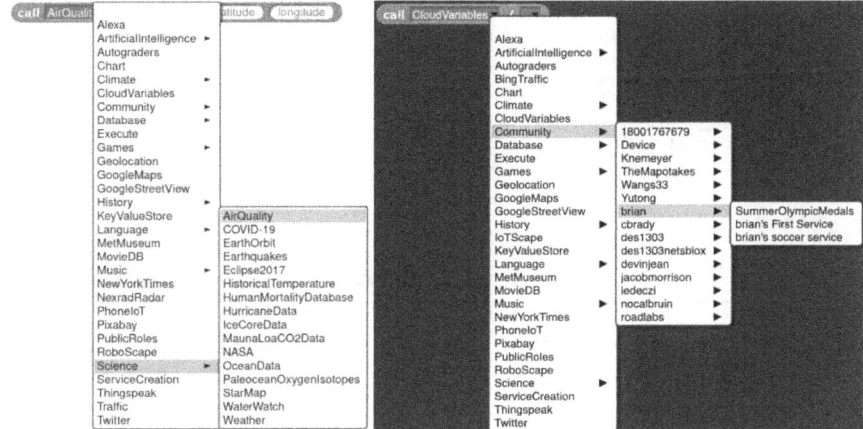

**FIGURE 4.3**    Various datasets and APIs accessible through NetsBlox (left). Service creation capabilities of NetsBlox to support student- or teacher-provided datasets (right)

and Chart plotting (Figure 4.3a). All of these services access real-world datasets that could be used in a myriad of DS or machine learning projects relevant to students. In addition, students and teachers can upload their own data for analysis and visualization as described below. Together, this easy access to an ever-growing set of databases and APIs affords what Resnick and Silverman (2005) would term "wide walls" to support a wide range of explorations to support varied student inquiry projects and interests.

*Create Your Own Service*

One of the most powerful features of NetsBlox for supporting authentic, student-driven, data inquiry is the ability to bring in any dataset for analysis and visualization, be it downloaded from the internet or from surveys conducted in the school, or cultural datasets from students' communities. Leveraging this feature can make for equitable and culturally relevant DS and learning of topics such as machine learning (Register & Ko, 2020). NetsBlox's "Service Creation" feature makes it possible to add one's own Service that will then appear in the "call" block pull-down menu under "Community" and then *username* submenus (Figure 4.3b). To do this, all one needs is a CSV file with the data that the Service will supply. The user can also simply drag and drop the file into the NetsBlox window to convert it into a variable (a two-dimensional array) automatically.

Figure 4.4 shows a simple example program that pulls in data from NSF's National Center for Science & Engineering Statistics or NCSES (https://www.ncses.nsf.gov/) database on new CS PhD students each year, by ethnicity. Such

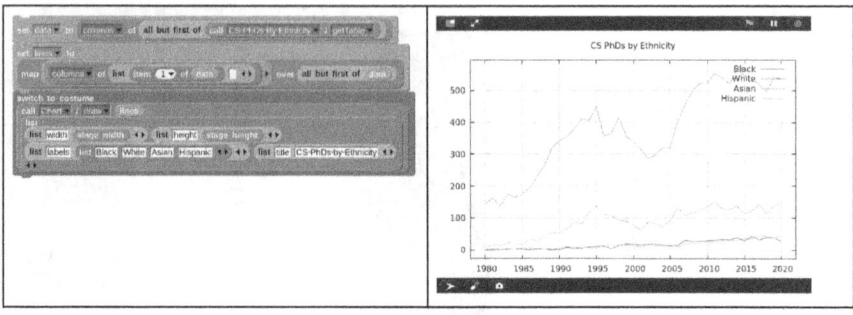

**FIGURE 4.4**   Code to plot the new CS PhDs data by ethnicity (left) and the resulting chart (right)

a service could be used (and perhaps extended by students) to examine the sociocultural and sociopolitical realities of race-based disparities in the field of CS supported by visualizations such as the one in Figure 4.4b.

### Engagement in Data Practices

Various NetxBlox capabilities support key data practices for a CS+DS curriculum, such as traversing and accessing elements of data structures, data visualization, and use of abstractions for pre-processing and transforming data into forms amenable to visualization and analysis.

### List Processing

Traversing a list/array or multidimensional data structure is key to DS-related coding. Because Snap! and NetsBlox introduce students to a variety of CS topics, they both provide several different real-world approaches to working with and operating on lists. This includes traditional index-based looping and more modern value based looping, as well as several functional programming techniques such as mapping a function over a sequence or combining elements of a sequence using a function. As yet another approach, Snap! and NetsBlox also support tensor-like operations on lists, which are extremely common in contemporary large-scale DS and machine learning. As a simple example of these different list processing paradigms, the following four code snippets all compute the sum of squares of numbers in the input "values" list (Figure 4.5).

### Data Visualization

Visualizing data in a graphical format is a key to DS. It allows sense-making with data and aids the process of interpreting and analyzing data through visual representations.

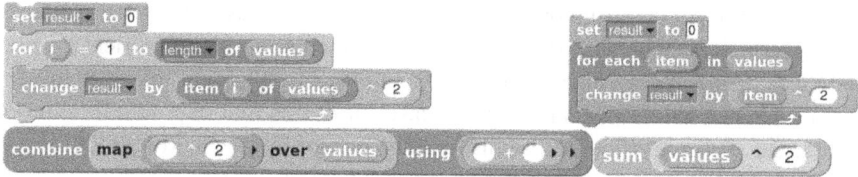

FIGURE 4.5    Four different code snippets that achieve the same goal (to compute the sum of squares of numbers in the input "values" list) demonstrate the different list (and data) processing paradigms available in NetsBlox

Snap! and NetsBlox are natively graphical environments, including a number of features such as sprites, costumes, and pen trails, which many Snap! students use to build games. NetsBlox, with its more data-focused curriculum, leverages these features for convenient data visualization. For instance, a sprite's appearance is controlled by its costume, which can be any arbitrary image. Because of this, NetsBlox includes web-based services (such as Google Maps or Chart) that return images that can be displayed by sprites or the stage by setting them as its costume. One of the most useful NetsBlox utilities for data visualization is the Gnuplot-based Chart service, which can be used to plot 1D or 2D data in various styles such as lines, scatter plots, bar graphs, and so on.

Figure 4.2a demonstrates how a simple plot easily visualized the steadily rising levels of $CO_2$ in the earth's atmosphere over the last 60 years. While climate change is a critical issue of great interest to young students, day-to-day issues of weather and air quality also affect students' lives and communities. Both issues have health equity implications, as they impact the poor and communities of color disproportionately (Bell and Ebisu, 2012; Fagliano and Diez Roux, 2018). Figure 4.6 shows an example of a NetsBlox project where the stage displays a Google Maps image in the background while a sprite queries points on the map for air quality information and uses it to construct a heatmap overlay. Culturally

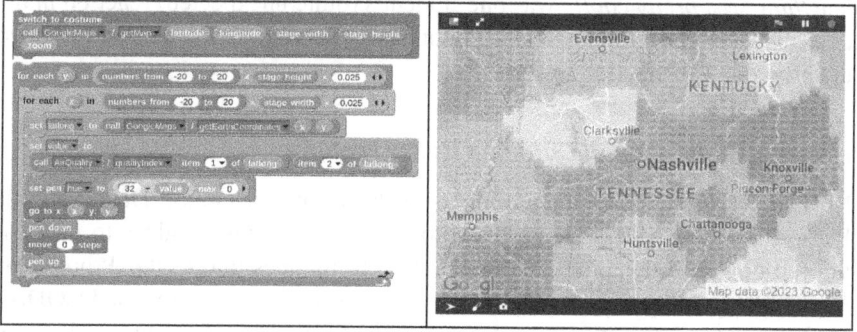

FIGURE 4.6    Air quality heatmap visualization overlaid on a map

relevant projects could involve students identifying areas with low air quality and overlaying them with data on household income, population density, and ethnicity. Students can also create more complex visualizations of other topical data such as Covid-19 vaccinations or cases or deaths that require a little more work and pre-processing before they can be passed to the Chart service, as described below.

### Pre-processing of Data and Easy-To-Use Abstractions to Support Complex Data Manipulation and Data Visualization

Litts et al. (2021) argue, "*Designers of learning technology challenge us to design beyond these 'black box' instruments by making tools that are at the right 'level' of understanding, making key concepts visible as appropriate to learning goals* (Resnick, Berg, and Eisenberg, 2000)". The beauty of SnapI and, by extension, NetsBlox is that custom blocks allow you to "black box" complex procedures while at the same time allowing them to be opened by the curious student who wishes to lift the hood on pre-programmed abstractions. This "glass-boxing" of code is a convenient strategy when data returned from services require pre-processing to transform them into a form that is amenable to visualization/computation. When such data wrangling is deemed too complex for our target age group, it is easy to pre-program blocks and provide them to students in starter projects.

The Covid-19 pandemic presented several opportunities to harness DS and CS for interrogation of a topical and significant scientific and public health phenomenon that affected us all and also has equity implications (Lee and Grapin, 2022). However, the cases and deaths data from the JHU Covid-19 database are returned as cumulative figures. Plotting daily averages or moving weekly averages requires extracting that data from cumulative figures. In our pilot projects with students, we provided them with a "process data" block to accomplish these complex data transformations as shown in Figure 4.7. "*The moving average thing was pretty cool!*" was one student's (quite unexpected) reaction in the post-survey in the Climate Science pilot camp (Section 3.1) where we provided the pre-coded abstraction to students.

### Dynamic Exploration of Data and Its Key Features

NetsBlox also supports integration with the Common Online Data Analysis Platform, or CODAP (Finzer, 2016) to facilitate gaining insight into data using the powerful dynamic data exploration techniques that CODAP provides. Students are able to import datasets from NetsBlox into an embedded CODAP environment to interactively explore trends and patterns in the dataset. They can then iteratively process segments of the dataset discovered using CODAP in

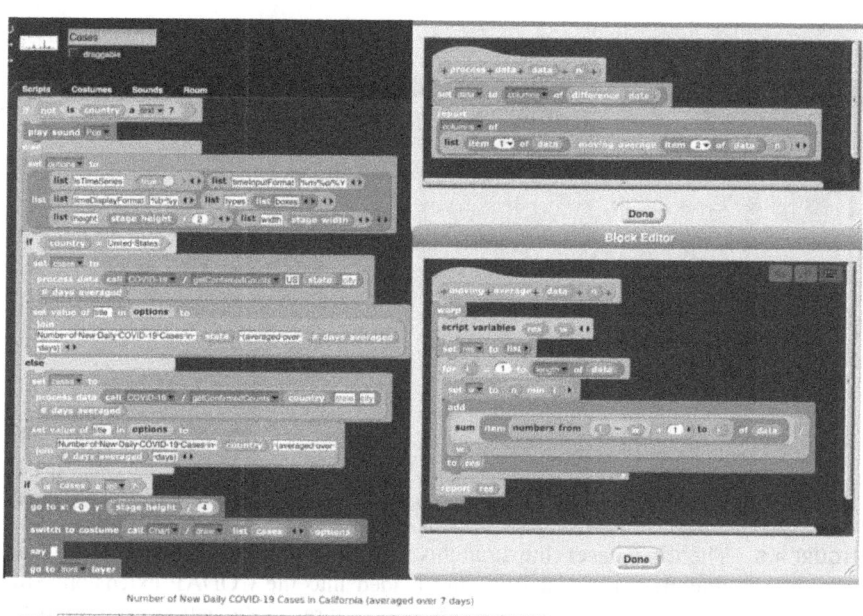

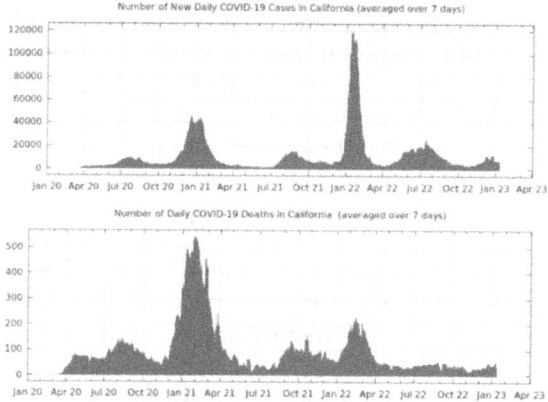

**FIGURE 4.7** Visualizing weekly moving average data of Covid-19 cases and deaths in California. "Process data" and "moving average" "difference" are preprogrammed abstractions provided to students

the block-based programming environment. This is particularly powerful in the context of understanding the impact of data on the training of machine learning models. NetsBlox not only embeds the CODAP UI but also allows selected tables to be pulled into the programming environment as multidimensional lists for subsequent coding (Figure 4.8). In a related NSF effort focused on AI and Cybersecurity (DGE #2113803) led by Grover and Broll, learners use CODAP to explore datasets before building ML classifiers in NetsBlox.

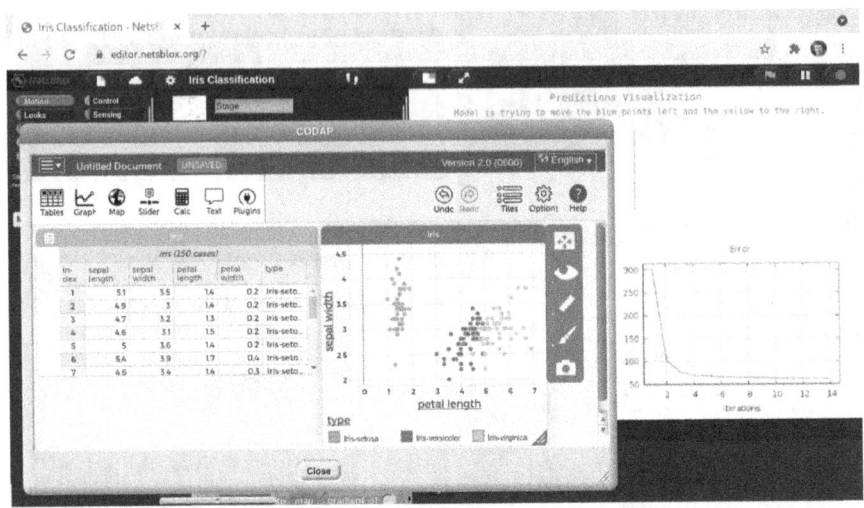

**FIGURE 4.8** The iris dataset (https://archive.Ics.Uci.edu/ml/datasets/Iris/) has been imported into NetsBlox and loaded into the CODAP extension. Students are able to explore different features of the dataset, such as petal length and sepal width, and how they relate to the data labels. A student may notice that the petal length is a simple indicator of the iris-setosa plant. They can then select the remaining points to process (if desired) in NetsBlox and explore them further using CODAP

### Direct and Personal Connections to DS through "PhoneIoT"

Recent additions to NetsBlox services allow students to connect their mobile devices to NetsBlox and access live data from their multitude of phone sensors. This facilitates a wide breadth of projects that give students a direct and personal connection to DS projects (Jean et al., 2023). Since the phone is an easily accessible personal device, NetsBlox's *PhoneIoT* feature (that drives the CSF Internet of Things module) provides excellent opportunities to promote creative, authentic engagement and supports projects that involve personal data collection and analysis through wearable activity trackers (Lee, 2018). For example, the phone's dedicated step counter sensor opens up a world of possibilities for "quantified self" DS projects. The creation of such personally and culturally relevant projects and artifacts position students as both the collectors of data and the agents to which the data references, and the data collection also interacts with the cultural milieu within which the data are collected (Lee, Wilkerson, and Lanouette, 2021).

Additionally, the Phone IoT interface is *interactive and customizable*. Unlike other tools which allow for reading sensor values remotely in real time but restrict users to only using the device as a sensor hub (in addition to often being complicated and expensive due to running in fee-based, specialized

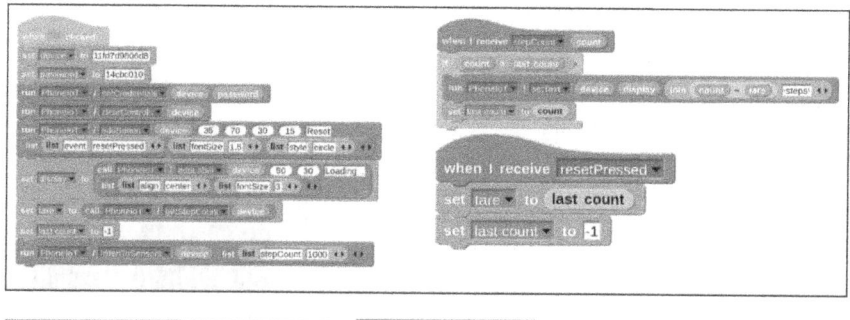

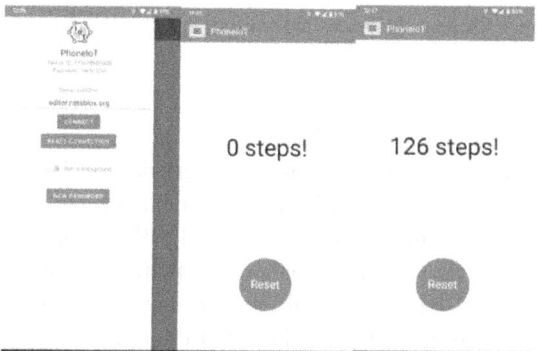

0 steps!          126 steps!

**FIGURE 4.9**   An example of using both PhoneIoT's sensor-based and graphical-based features. The program converts the phone into a step counter that tracks the user's number of steps in real-time and updates the app's display with the relevant information. The program also has a "reset" button to the display which, when pressed, sends a message back to the NetsBlox project (running in the browser) to zero out the number of steps

cloud environments), PhoneIoT promotes interactivity through a customizable phone display, which allows teachers to cover important topics like event-based graphical inputs and remote controllers (see Figure 4.9). Lastly, PhoneIoT also helps *broaden participation through physical computing and interdisciplinary connections.* Recent research highlights several benefits associated with physical computing including increased motivation for students, those from diverse backgrounds, because working with sensors is tangible and often practical (Sentance et al., 2017). The *PhoneIoT* app is freely available on both Android and iOS, and is designed to cover as many different phone models and versions as possible.

### Transition to Text-Based Data Science Related Coding in Python

Although NetsBlox supports engagement in authentic DS inquiry through black-based programming, the reality is that the typical programming languages used

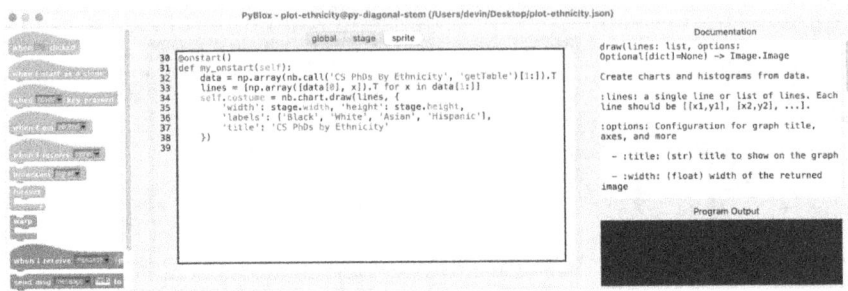

**FIGURE 4.10** PyBlox provides an easy block-to-text transition from NetsBlox to Python

for real-world DS are textual, with Python being the most popular. To create a gentle bridge for students from introductory block-based experiences in Nets-Blox to Python, we have developed *PyBlox,* which allows students to transfer much of their existing NetsBlox knowledge into native Python. PyBlox supports all the same features such as RPCs, as well as graphical components such as sprites, which means that students do not have to rethink how they write code, and can focus on learning Python. Additionally any installed Python packages can be imported into PyBlox scripts. Figure 4.10 shows an example of using the PyBlox IDE with the popular DS tools, NumPy, to implement the same block-based data visualization project shown in Figure 4.4.

## Pedagogical Underpinnings of CS Frontiers Curriculum Design

**Humanistic engagement with data.** Lee et al. (2021) articulate a framework for promoting equity in DS education by taking a humanistic approach. Their humanistic DS framework aligns well with Ladson-Billings's (2014) *culturally relevant pedagogy* and suggests that students engage with data and datasets at three levels: personal ("students' personal and direct experiences with data, measurement, and the contexts in which data are collected"), cultural ("the cultural and sociotechnical infrastructures and values enacted during a dataset's collection and use"), and sociopolitical ("the enduring political and social narratives that affect the purposes and methods by which datasets are constructed, interpreted, and used as social texts"). Through projects pertaining to climate change, healthcare, social media, music and movies, social justice, and their impacts on communities, the spectrum of CSF curricular activities in NetsBlox ensure opportunities for working with datasets pertinent to the personal, cultural, and sociopolitical.

    **Collaboration.** All CSF activities are designed to be done collaboratively. Research suggests that girls prefer collaborative environments, which can help boost self-confidence and foster a sense of community between girls (Pollock et al., 2004). In our curricular design, we intentionally created activities that specifically require collaboration, not just account for it as an afterthought. Supporting equitable collaboration is imperative such that individual students do not feel ignored, belittled, or not useful for a project (Lewis and Shah, 2015). We take inspiration from research by "the programmers collective" which stresses the importance of "multiple levels of designed-for collaboration" (Fields, Vasudevan, and Kafai, 2015). As more schools begin adopting CS or coding experiences, we must acknowledge that students will enter a classroom with large variances in their coding or computing knowledge. Preparatory privilege often gives more experienced students a sense of entitlement and the initiative to take lead (Margolis et al., 2008; Margolis and Fisher, 2002). Unfortunately, this leaves the more inexperienced students, disproportionately girls and minorities, in a position of lower power and consequently less active participation (ibid). Netsblox affords users the opportunity to program simultaneously and on varying components of the project, or even in a jigsaw manner where each student can use different subsets of blocks (Lytle et al., 2020). The choice of a culturally relevant software design task (say, media production) can encourage complex programming and invite coordinated collaboration, as can distributing the tasks so individuals feel ownership in their own parts and of the collective whole.

**Scaffolded project-based activities.** PBL activities that are collaborative and connect to varied everyday topics of interest (e.g. healthcare, climate change) have demonstrated potential for engaging young women and underrepresented minorities in computing and STEM (Estrada et al., 2016; Fields and Kafai, 2020; Lee, 2020). Expansive-framed curricular designs that make relevant, real-world connections to students' lives and futures also promote interest and deeper connections for transfer (Engle et al., 2012; Grover, 2021c).

Furthermore, as high school students come to CS+DS courses with varying degrees of expertise, it is imperative that we support all levels of prior preparation, lifting up those that are just getting started and widening the horizon for those who have been exposed to computing for much longer (Resnick and Silverman, 2005). This philosophy also considers the teacher's role in the classroom as the facilitator, and that having students at varying knowledge levels can be quite overwhelming. Instead of moving advanced students further along into the curriculum we pose the idea of letting students choose extensions that enhance the overall project, but do not require additional coding skills.

Additionally, CSF modules combine PBL with "Use-Modify-Choose" (Lytle, et al. 2019a) an adaptation of the familiar "Use-Modify-Create" framework (Lee et al., 2011). These curricular designs promote confidence in coding by careful sequencing, so that students and teachers first use code, then solve code puzzles that require small modifications, and then choose what to make based on their own interests (Lytle et al. 2019a). The "Use" phase of the lesson lets students become familiar with the environment by trying out an example and making connections to an unplugged activity. This puts students on equal footing, as the main goal of the task is exploration and building connections. It also allows teachers to build confidence in their own understanding of the program that all the students are learning. In the "Modify" phase, students begin to act on their understandings and can edit the provided projects and complete scaffolded examples (Lytle et al., 2019b). We often employ pseudocode and/or Parson's puzzles (Parsons and Haden, 2006) where students are presented with jumbled-up blocks of code that they must assemble in the correct sequence. Parson's problems are growing in popularity as a pedagogical tool that has demonstrated benefits in introductory programming settings (Ericson et al., 2022; Zhi et al., 2019). In each of these phases students are given increasingly challenging tasks, however, they are still scaffolded to stay within a certain level of comfort (as in, e.g. Repenning et al., 2015). The "Choose" phase addresses an issue that novice students often face– the blank page syndrome, which is not knowing where to get started when told simply to "make something" (Limke et al., 2022). In this Choose phase, students are given choices of potential extensions or new "proximal" projects with other datasets that allow them to follow their interests while still building on familiar concepts in their own ways. For the teacher, the list of choices helps them be prepared to scaffold students' explorations. One of the choices always is, of course, doing a project totally of one's choosing.

As an example, a project investigating invasive species through computational modeling in NetsBlox might give students the base project of predicting whether or not Kudzu is invasive in a particular area given some EPA data. Students would then program the ability to click on a map of a particular region to see whether or not Kudzu is prevalent in that area based on EPA data. This base project connects to environmental studies, data interpretation, and also supports collaboration through determining prediction algorithms, as well as varied code tasks (map interfacing vs algorithm design) which reflect different roles in industry. Furthermore, Choose extensions could allow students to focus more on the science side and investigate other invasive species, make a prediction game to see if the user's prediction matches the data, or focus on math or visualization and add charts to show the Kudzu-infested regions. These extensions allow students to self-assess their own abilities and interests and apply them to the lessons to feel ownership not only of the base code, but of the new programs that each look different than those of their peers. Such rich and diverse extension opportunities, along with the collaboration support afforded by NetsBlox (as described below), may improve the perceptions, confidence, and performance of girls and minorities. Furthermore, the explicit instruction of software engineering techniques and engineering design principles can help students learn common practices that are used within the information technology industry (Gransbury et al., 2023).

In the following sections, we describe two curricular interventions that showcase how these design principles come to life.

### Climate Science Pilot: Data as a Vehicle to Deeper Understanding

Research suggests that bridging the gender gap in computing requires presenting CS as a variety of perspectives and possibilities (Khan and Luxton-Reilly, 2016) and making connections to other subjects (Margolis and Fisher, 2002). Numerous studies suggest that climate change is the leading real-world issue that concerns our students and youth. Student engagement with climate science curricula has also been shown to promote responsible personal choices (Cordero, Centeno, and Todd, 2020). The importance of climate science is also emphasized in the NGSS (Lead States, 2013), as it cuts across several disciplines in the natural and social sciences (Hestness et al., 2014). Additionally— and importantly—a wealth of climate and paleoclimate data are made available online through services such as the National Climate Data Center (https://www.ncdc.noaa.gov/cdo-web/), providing an exciting opportunity for students to explore and understand the science of climate change on multiple time-scales, provided that they have the necessary tools to understand, visualize, and engage with the data. Such experiences that focus on science practices also invite teachers to attend closely to the varied ways in which students argue from evidence or interpret data as a foundation of learning in science (Bang et al., 2017).

The "Climate Change and Computing" (CCC) mini-course was designed to engage students in the science of climate but also in the attendant analyses of data and trends to understand the climate crisis better. CCC introduced students to NetsBlox's easy-to-use cloud and distributed computing features to access data from real climate data sources. CCC prompted students to "analyze geoscience data and the results from global climate models to make an evidence-based forecast of the current rate of global or regional climate change and associated future impacts to Earth's systems", thus directly addressing NGSS Standard HS-ESS3-5 (Lead States, 2013), while also engaging in an authentic interdisciplinary experience.

CC brought together a team of learning scientists, computer scientists, and climate scientists to facilitate an experience involving climate science, CS, and DS in an authentic interdisciplinary setting. Members of the CSF team (Ledeczi, Broll, and Grover) collaborated with Dr. Jessica Oster, a paleoclimatologist and faculty at the Vanderbilt School of Earth and Environmental Science, who led the development of a series of "CS+DS and Climate Science" lessons and group activities that

1  Provided students background information on climate science, including the tools scientists use to understand past and present climate and make informed predictions about the future;

2 Promote student engagement and supported interpretation of real climate data through visualizations and computations completed in NetsBlox; and

3 Promote open-ended inquiry and introduce students to research question development which is central to the scientific process.

**CS+DS and Climate Science Camp:** Over a period of eight weeks, a virtual camp was offered to 26 grade 9 high school students in 2021. The camp was delivered virtually due to the Covid-19 pandemic. The camp participants were majority females (69%) and self-identified as Hispanic or Latino (35%), White (31%), Asian (15%), Black (8%), and Other (12%). The students were drawn from nine different high schools that spanned geographically and demographically diverse regions across Tennessee. Due to the pandemic, activities were reduced to a weekly one-hour online class on Zoom. Since we could not assume previous programming background (indeed, some students had none), introductory activities involved (1) an introduction to programming using NetsBlox through a Weather app that introduced students to variables, branching, custom blocks (functions), and RPCs and (2) a scaffolded homework that involved creating a project that utilized the Movies Service to display the photos of the leading cast members of any movie based on the user provided the title. In weeks 2 through 4, the time was split evenly between CS+DS and climate science as shown in Table 4.1. The CS+DS focus was on data processing and visualization. The second half of the camp was dedicated to team projects.

With regard to climate science, after a lesson on climate and paleoclimate science basics, students were encouraged to interpret and discuss time-series plots of global temperature anomalies that incorporated data from several paleoclimate archives as well as model output for future

**TABLE 4.1** Computer science+data science and climate science curriculum (over 8 weeks)

| Computer Science and Data Science | Climate Science |
| --- | --- |
| Introduction to NetsBIox and Distributed Computing (Weather app) | N/A |
| Intro to two-dimensional arrays and plotting. Plotting $CO_2$ from Mauna Loa | Intro to paleoclimate information, Temperature, $CO_2$, carbon cycle |
| Plotting ice core data | Ice core data, d18O proxy |
| Data processing. Computing mean and moving average | Review of proxy information, discussion and plot interpretation, Q&A |
| Project assignment | Project assignment |
| Project work and help | Project work and help |
| Project work and help | Project work and help |
| Project presentations | Project presentations |

climate change scenarios. This exercise provided practice with graph interpretation and allowed the students to consider the qualities of an effective data visualization graphic. Data analysis in the service of sense-making in science also motivated the use of data practices and data visualization, working with multidimensional datasets (implemented as multidimensional data structures in code) and statistical concepts such as mean and moving average which was provided to students as a pre-coded block (abstraction) as described in Section 2.3.3.

For the final project, the six teams were assigned one of three topics each associated with climate/paleoclimate datasets from the scientific literature. These included air temperature proxy data from South American mountain glacier ice cores [28], sea surface temperature proxy data from corals from the Yucatán [30], and global air temperature anomaly and climate forcing data compiled for the 2018 National Climate Assessment (https://nca2018.globalchange.gov). Students were provided the original references for their datasets as well as other resources to learn more about their kind of data. Each group was also given a set of questions to guide their research into each dataset, but beyond that, the assignment was intentionally open-ended. To support the students, there were discussions focused on identifying the qualities of good scientific research questions. They were to use NetsBlox to process and visualize the data they needed to answer their research questions.

The end product for each team was a presentation in which students described the data archive they were working with (corals, ice cores, climate forcings), talked through their dataset, analyses, and visualizations coded in NetsBlox, and ended with an original research question that could be pursued with their data. This structure ensured that each group pursued independent research and also asked that the students demonstrate the creativity and self-directedness required for scientific research (Figure 4.11). During the presentation, one student commented, "I think the most interesting thing I learned, mostly just because it was a Eureka moment, was that anthropogenic forcings have clearly caused the increase in average global temperature".

**Data and results:** Post-survey questions asked:

1  What were one or more takeaways, aha-s, insights regarding CS from this program? (Open response);
2  This program expanded my views regarding how we can use programming [Slider: Definitely No(0) to Definitely Yes(100)];
3  I am interested in learning more about how we can combine programming with other subjects/topics (such as climate change) [Slider: Definitely No to Definitely Yes]; and
4  What are some topics you'd like to explore through programming? (Open response).

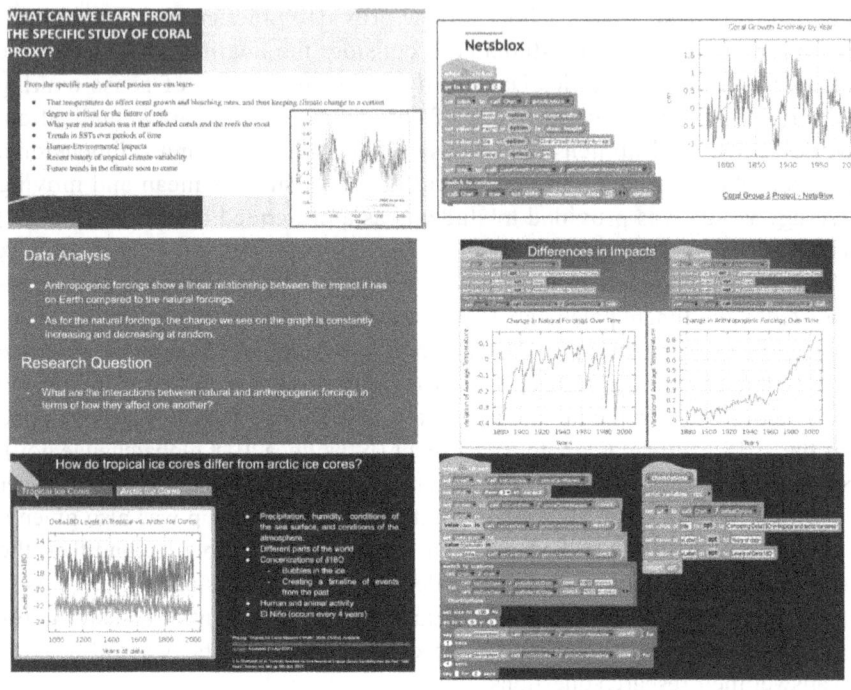

**FIGURE 4.11**    Snapshot of students' final climate science CS+DS project presentations

Several interesting findings emerged from the post-survey responses. Data indicated that the program expanded students' views regarding how they can use programming (83%) and that they were interested in learning more about how to combine programming with other subjects/topics, such as climate change (73%). The responses to Q1 were coded and fell in one of six categories (see Figure 4.12). Student responses to Q4 were overwhelmingly encouraging (Table 4.2). Close to 75% of responses made reference to topics that appeared to be built directly on the climate change experience. Many were about analyzing data from various aspects related to human and animal existence, including health, psychology, animal extinction, disease, and vaccines. These responses and other results suggest how data drawn from, and relevant to answering questions on, climate change motivated organic engagement in data practices and visualizations. In the process, it also fostered positive perceptions of—and generated interest in—computing among the majority of female camp participants (Grover et al., 2022). Such engagement with data fosters equity in STEM learning and also prepares learners for new data-intensive topics such as AI. This effort shines a light on the affordances of NetsBlox and its simple yet powerful abstractions that made this experience possible for high school students (not all of whom had prior programming experience).

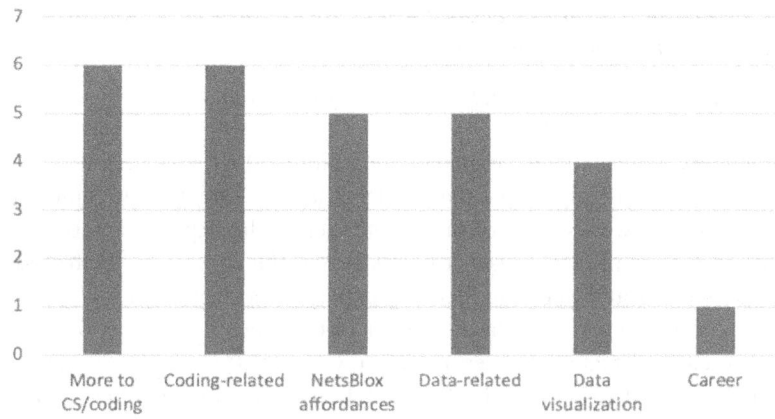

**FIGURE 4.12**   Main themes in student responses to the question "What were one or more takeaways, aha-s, insights regarding computer science from this program?" After the **CS+DS & Climate Science Camp**

**TABLE 4.2** Responses to what are some other topics you'd like to explore through programming?

---

The worlds chain in the food industry that include animals. Like fishing, whale killings, tuna killings and all that. See on average how many we are killing for our own usages and stuff I think that would be really cool because I have seen a lot of documentaries on it and I want to collect data on it or some sort of that.

I would like to explore something about population levels in different ecosystems. I have always been very interested in animals and their environment and think that it would be cool to try seeing different data with computer science (CS) to understand the effects of climate change on ecosystems.

I would like to look at the changes in the plastic in the ocean over a period of time.

I would like to research more about the health of humans and what affect humans the most.

I would like to explore how to use art, programing, and STEM ideas for a future career space and physics.

I would like to explore how we can use CS to analyze human habits and psychological patterns more.

Air quality

I would like to explore ratios of vaccines and diseases using graphs.

I'd also like to learn about how programming relates to electronics and how we can use it to direct electricity.

Another topic I would like to explore through programming is the solar system.

Maybe like baseball batting averages or the WAR or Wrc+ or in football to compare QBR and to analyze data in sports using coding.

I'd like to explore extinction/decrease in animal populations.

---

### Ethics, Equity, and Social Relevance – Socially Relevant AI Curriculum; Data and Findings from Two Pilots

*AI applications can impact society in both positive and negative ways. (Big Idea #5,* Touretzky et al. (2019)

Given how new advances in machine learning and AI have serious implications for equity, it is imperative that students in K-12 are able to explore how machine learning works. One of the advantages of the capabilities for accessing web services and real datasets of NetsBlox is the ease with which students can engage in AI and machine learning experiences. NetsBlox not only provides access to any and all datasets (either through pro-defined services or the "service creation" feature, as described above), but it also provides access to APIs such as Parallel Dots for NLP and text analysis with a simple "call". For example, the call/ParallelDots/getSentiment/text block returns the overall sentiment of the "text" along with a confidence score. The returned data structure has a confidence level for each of the sentiment categories: negative, neutral, and positive.

Leveraging these low-floor and data/API capabilities of NetsBlox, we designed an AI/ML module that focused on ethics, equity, and social relevance and was designed for generating and promoting interest among high school girls. We were guided by the AI4K12 Big Idea #5—high school students should be able to evaluate new AI technologies and describe the ethical or societal impact questions raised by them (Touretzky et al., 2019) as well as scholarship on engaging females in computing. Additionally, connecting real-world experiences that make an impact with diverse female experts for support and inspiration can provide girls with authentic STEM opportunities that promote sustained engagement (Chapman and Vivian, 2017).

### Artificial Intelligence Camps

During the summers of 2021 and 2022, we offered week-long virtual camps to a total of 34 high school students. Both camps were delivered virtually due to the Covid-19 pandemic. The camps were conducted by high school CS teachers (three females and one male; two Black, and two White) who received professional development. Teacher participants suggested ways to improve activities and collaborated to adapt the instructional materials for curricular implementation. They then co-taught the camp in pairs (camp student participants were divided into two groups). Over half of the camp participants were females (56%) and self-identified as Asian (65%), White (15%), Black (9%), and Other (11%). The students were recruited from several different high schools in regions across Tennessee and North Carolina. Each camp was one week long (five weekdays), and each day, activities and lessons took place over seven hours (that included time for lunch and breaks). Students were required to have some prior programming experience, but experience in a block-based programming environment was not a prerequisite. Because of this, there was an introductory activity during the first hours of day 1 for students to gain familiarity with the NetsBlox environment. In days 2–4, time was split between instructional activities (including unplugged and programming activities), explorations of freely accessible online tools (such as *QuickDraw*) and extension programming activities (in NetsBlox and Python). Some programming activities were introduced through pseudocode and/or Parson's Problems. See Table 4.3 for the AI/ML camp curriculum outline.

### Engagement with Data and Exploration of Ethics and Bias

Early in the curriculum, students are introduced to ideas of data and machine learning while also engaging them in the interrogation of bias and ethics in AI and data analysis as a throughline in all the activities. The AI & Drawing activity involves explorations of a program that uses ML, the data that trained that program, as well as issues of ethics and bias. Students play with Google's *Quick,*

**TABLE 4.3** Artificial intelligence and programming concepts curriculum (over five, seven-hour days)

| Day | Lesson | Programming Activities |
|---|---|---|
| 1 | Introduction to AI and algorithms | Ranking AI |
|   | Programming concepts | Conditionals |
| 2 | AI and drawing | Quick, Draw! |
|   | Programming concepts | Strings |
|   |   | While Loops |
|   |   | Lists |
|   |   | Advanced Decision Tree Structures |
|   | Breadth First Search | Breadth first search and social media analysis exercises |
| 3 | Classification | Twitter Bot Detection |
|   | Natural language processing and sentiment analysis | Sentimental Writer |
|   |   | Sentimental Writer and choose your own adaptation |
| 4 | introduction | Music Sentiment Parson's Problem |
|   | Neural networks | Introduction to Imitation Learning |
| 5 | Ethics | How AI effects our lives |
|   |   | AI and ethics |
|   | Real-world applications | AI and criminal justice |
|   |   | AI and environment |

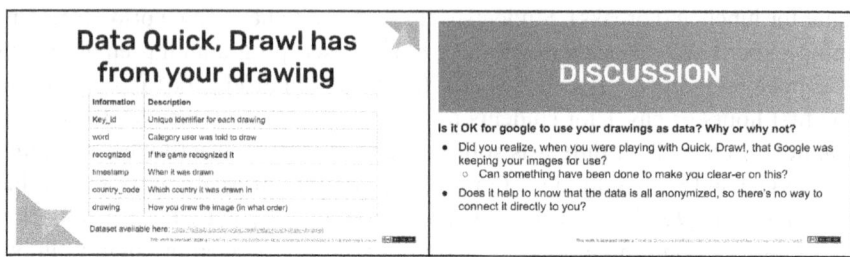

FIGURE 4.13    CS Frontiers (CSF) slides to engage students in a discussion of user data collected and the ethics of training ML models like Google's *Quick, Draw!*

*Draw!* app. They examine the datasets and watch a video that explains how gathering millions of drawings helped Google train its ML models. Students examine how gendered biases crept in the training of *Quick, Draw!* through watching a video and discussing how such bias could be mitigated or avoided in the first place. Students also interrogate the ethics of Google created an app like *Quick, Draw!* They made aware of the kinds of data Google gathered from users without their knowledge, and discuss the ethics of such data collection activities from unsuspecting users (Figure 4.13). They also discuss Cathy O'Neill's short video on *The Truth About Algorithms*. In addition to explorations with freely available AI tools, activities in the "Ethics and Real-world Applications" unit were dedicated to examining and discussing examples of how the bias of training data in machine learning (ML) programs can cause bias in decision-making the ML model is being used for. Examples included facial recognition, the role of AI in the criminal justice system, and how environmentalists use AI. Students interrogated environmental issues and "AI for good" to help people and animals during the AI & Environment activity, as well as pernicious and potentially harmful uses of machine learning trained on historical data that mirror societal biases and racist policies in the criminal justice system.

Students also engage in other data experiences connected to relatable issues in the real-world through the Twitter Bot Detection, Sentimental Writer, and Music Lyrics Sentiment activities (Figure 4.14). The purpose of the Twitter Bot Detection activity was for students to understand the concepts behind machine learning data classifiers. At the beginning of the activity, students completed an unplugged exercise where they were asked to examine a dataset of Twitter account metadata such as the number of followers, active duration of the account, and so on. From this dataset, students were asked to identify which accounts they thought were maybe bots, i.e. computer programs acting as application users. After this exercise, students then created a data classifier to detect which accounts were bots within the dataset. Students realized that the more specific

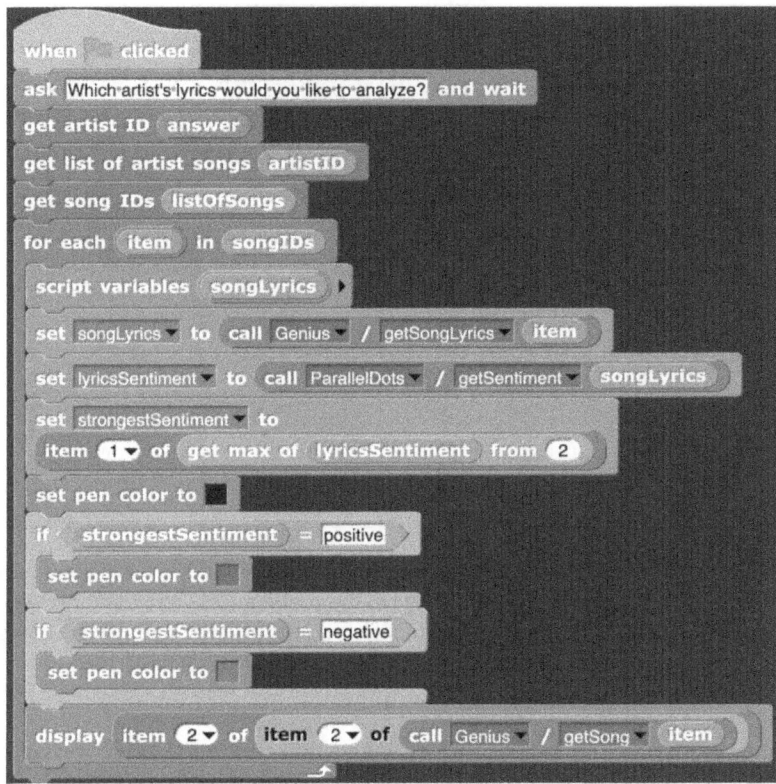

**FIGURE 4.14** Music Lyrics sentiment analysis using the ParallelDots NLP text analysis API and the Genius data library for song Lyrics.

they were when making conditions in the classification algorithm, the more successful their program was at detecting bots.

Students also completed the Sentimental Writer activity using Python in 2021 and NetsBlox in 2022. We chose to change the language to NetsBlox because of the time constraint of the one-week summer camp, but the teacher facilitators (who taught both in 2021 and 2022) indicated a desire to use Python as well if students had more time with the curriculum in the classroom. The purpose of this activity was for students to use the Parallel Dots NLP API to determine the sentiment of text from New York Times articles, analyze the results, and deliberate on the limitations of the AI (NLP) Tool (Figure 4.15). Students first programmed an input box that allowed them to test the sentiment of various phases and words they input into the program to see how the analysis tool categorized specific language. Next, they analyzed select New York Times articles of their choice. In an extension of the activity, students were able to detect other attributes such as sarcasm or abusive language. Many students found this to be

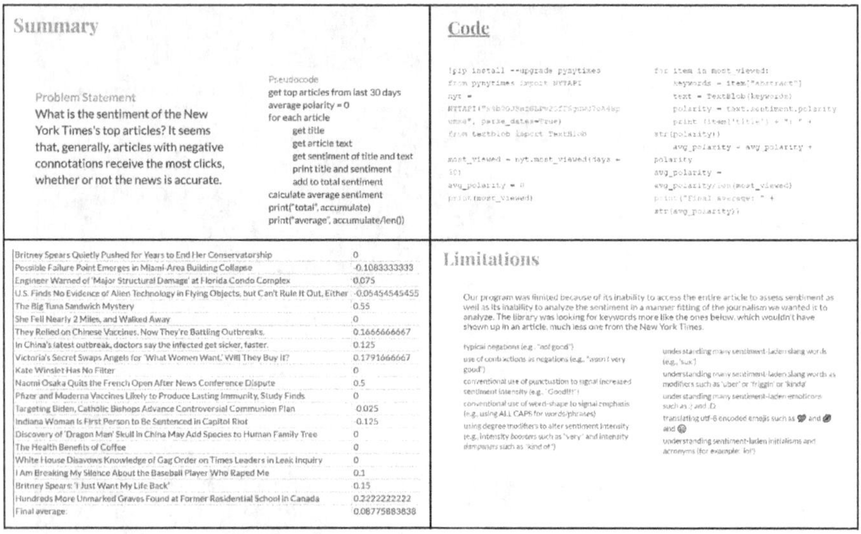

FIGURE 4.15    Presentation by two female students of their NYT articles sentiment analysis project including (a) the problem statement and pseudocode, (b) Python code, (c) results for various articles, and (d) an analysis of results and limitations of the sentiment analysis tool

an interesting activity because of its connection to current issues around hate speech and bias in journalism occurring across the globe.

To connect curricular activities to students' interests, students extended learning from the Sentiment Writer to the Music Lyrics Sentiment analysis activity where students analyzed the lyrics of their favorite artist's songs. First, students were asked to develop a hypothesis about how the sentiment of an artist's song would be classified. Then, they completed a Parson's problem to create a program that accessed song lyrics data from the Genius library and analyzed artists' songs (Figure 4.15). After a discussion about whether their hypotheses were rejected or not, students learned that when examining data, one has to account for one's own biases and have a substantial amount of data to make generalizations about a topic.

## Data and Results

We administered pre-surveys and post-surveys to assess students' confidence on items related to AI and ML content, career identity, and self-efficacy in CS. Table 4.4 shares the results of female participants' responses in the Summer 2021 pilot study. Q15 shows a significant difference $t(9) = 3.28$,

**TABLE 4.4** *T*-test results showing the mean difference in female participant confidence from pre-survey/post-survey from pilot summer Camp in summer 2021 (alpha = 0.018) (*N* = 10)

| Survey Question | Pre-test Mean | Post-test Mean | MD | SD | t | P |
|---|---|---|---|---|---|---|
| Q.15: Right now, how confident are you in your ability to understand how computers present data and images? | 2.00 | 2.70 | 0.70 | 0.68 | 3.28 | 0.01* |
| Q.38: How much do you know about how to build artificial intelligence applications? | 1.40 | 2.30 | 0.90 | 0.74 | 3.86 | <\$0.01* |
| Q.42: I am able to do well in activities that involve computer science (CS). | 2.90 | 3.70 | 0.80 | 1.03 | 2.45 | 0.04 |

$p = 0.01$ in female participants' confidence in their ability to represent data (and images). Their confidence increased from ($M = 2.0$, SD = 1.05) to ($M = 2.7$, SD = 1.05) on the four-point Likert scale. The second question of interest, Q36 measures participants' confidence in how much they know about building AI applications. This questions shows a significant difference $t(9) = 3.86, p < 0.01$ in participants' confidence that increased from ($M = 1.40$, SD = 0.70) to ($M = 2.30$, SD = 0.82). Finally, Q42, shows a significant difference $t(9) = 2.45$, $p = 0.04$, from pre ($M = 2.90$, SD = 0.99) to post ($M = 3.70$, SD = 0.95) with an increase in confidence in completing CS activities for female participants.

To gain insight into the female participants' camp experiences, we looked at open-ended questions from the post-survey on participants' experience in the camp and their interest in a career in CS. Four female participants commented on the impact of the female facilitators on their experience, confirming prior research that shows that role models can be another factor for female engagement (Margolis and Fisher, 2002; Sun and Clarke-Midura, 2022). Other responses are displayed in Table 4.5.

The sample size for the pilot camp was low ($N = 10$) in Summer 2022, and no significant differences were measured in participants' Likert-scale responses from the pre-survey to the post-survey. This kind of ceiling effect is often noted in STEM camps that student self-select into. However, we also gathered data on open-ended questions that probed participants' interest in continuing to learn about CS, what topics they may be interested in, and what their "takeaways" from the camp were. In Table 4.6, we share some of their positive responses.

**TABLE 4.5** Female camp participant responses to open-ended questions in the post-survey

I learned a lot and would recommend it to others.
I like how they were trying to make the students very energetic and more involving with the lesson' and 'I had fun at this camp.
I really enjoyed the camp and the support my teachers provided.
The information was very thorough and I really liked the step-by-step coding instructions. The Genius API activity [Music Sentiment activity] was also something I've never seen before (in a good way).
I liked how knowledgeable all the counselors were and the amount of effort that was put into the presentations and activity planning was evident.

**TABLE 4.6** Student positive responses

Gave me more insight on ethical concerns about computer science (CS)
AI week has changed my view of CS by showing me how ethical concerns and bias are in programs.
I am able to see real-world applications of AI.
I learned a lot about how AI is used for criminal justice (in a good and bad way).
I would like to explore AI and ethics more. I think it is an interesting cross section that results in interesting conversations.

**Year-long CSF pilot in AY2022-2023 as a high school course**. One of the CSF teacher partners piloted the full CSF course in the 2022–2023 academic year in a high school in Nashville, Tennessee. The CS teacher, who is Black and female, had many additional insights to share on her experience teaching the four CSF modules over nine weeks each in a mixed grade high school classroom. It is worth noting that five of six female students in the class of 13 who had no prior programming experience found the course to be interesting and engaging and one they would recommend to their friends (Brock et al., 2023). Specific to the AI/ML module, students extended the *Quick, Draw!* activity to examine gender bias through additional examples beyond the footwear example shared in the curriculum (handbags, for example). Furthermore, the release of ChatGPT generated a lot of excitement among the students for the AI/ML unit. Students created chatbots for their final project of the AI/ML module, for which they also gathered and cleaned their datasets. The teacher's feedback on the AI/ML unit echoed the positive sentiments from the pilot summer camp. One of her a-ha's related to the student experience was that students are not always aware that there is bias in AI tools—"*<A female student> says she felt like the machine learning was beneficial. She didn't really know that data was biased. Like these are things that like until they tell you, like, you don't know the kids think like this,*

*like you didn't know it was biased? She's like, she didn't know that the informa-*
*tion that she was getting from chatbots, things like that, was biased"*.

## Reflections and Conclusion

Based on feedback on student and teacher feedback on CSF interventions, we
believe CSF succeeds in its goals of using feature-rich tools to engender authentic learning experiences that promote equitable and humanistic engagement
with integrated CS+DS learning at the high school level. Whether it was an
interdisciplinary examination of climate change, an activity to help students un-
cover issues of ethics and bias in machine learning, or first-hand experiences in
classifying bots on social media or negative sentiments in mainstream articles,
CSF modules powered by NetsBlox provided accessible, hand-on experiences
connected to topical, real-world issues of concern and interest to today's youth.
Our deliberative designs for data-centered engagement with computing concepts foregrounded interdisciplinarity, authenticity, and equity (with special attention to gender equity). CSF interventions demonstrate how to make advanced
CS topics and thorny issues of our time accessible to students through deliberate
thought to tools and curricular design that centers data. Experiences like the climate science and AI/ML units described also helped expand students' horizons
about computing and how to engage with topics meaningful to them through
data and computing, while also making them aware of increasingly critical issues such as bias in data and AI/ML.

Edsger W. Dijkstra (1982) famously said, "The tools we use have a profound
(and devious!) influence on our thinking habits, and, therefore, on our thinking
abilities" (p. 14). This truism has guided our approaches to tool and learning
designs aimed at supporting student learning of computing and DS in ways that
promote authentic engagement and equity. Our chapter details how the design of
learning environments and tools (specifically, NetsBlox), and their affordances,
support engagement in data inquiry and advanced computing topics in integrated
CS+DS learning environments. It also shares how the design of activities in the
CSF curriculum, especially the AI/ML module, promotes engagement with real
datasets to make sense of topical issues and critically examine issues of ethics and bias in data and AI-powered tools. Through the two research examples
drawn from our work on the NSF-funded ITEST project, CSF, we demonstrate
how programming environments can democratize access to sophisticated and
authentic data-driven interdisciplinary inquiry as well as learning of emerging
CS topics such as distributed computing, Internet of Things, AI, and machine
learning (in addition to others, such as cybersecurity and software engineering,
not addressed in this chapter) for secondary school students while still maintain-
ing the low-floor that block-based programming experience afford to novice
programmers.

## Acknowledgments

This material is based upon work supported by the National Science Foundation under grant numbers 1949472, 1949488, 1949492, and 2113803. We are also grateful to the contributions of CS Frontiers colleagues Marnie Hill, Gordon Stein, Lauren Alvarez, and Janet Brock.

## Note

1 For a more detailed description of NetsBlox, its message passing capabilities (not addressed here) and its many expansive features to make advanced computing accessible to novices, see Brady et al. (2022).

## References

Astrachan, O., Barnes, T., Garcia, D. D., Paul, J., Simon, B., & Snyder, L. (2011). CS principles: piloting a new course at national scale. In *Proceedings of the 42nd ACM technical symposium on computer science education* (pp. 397–398).

Bang, M., Brown, B., Calabrese Barton, A., Rosebery, A. S., & Warren, B. (2017). Toward more equitable learning in science. *Helping students make sense of the world using next generation science and engineering practices* (pp. 33–58). NSTA press.

Bell, M. L., & Ebisu, K. (2012). Environmental inequality in exposures to airborne particulate matter components in the United States. *Environmental Health Perspectives, 120*(12), 1699–1704.

Brady, C., Broll, B., Stein, G., Jean, D., Grover, S., Cateté, V., & Lédeczi, Á (2022). Block-based abstractions and expansive services to make advanced computing concepts accessible to novices. *Journal of Computer Languages, 73*, 101156.

Brock, J., Gransbury, I., Cateté, V., Barnes, T., Grover, S., & Ledeczi, A. (2023, August). Student Attitudes During the Pilot of the Computer Science Frontiers Course. In *Proceedings of the 2023 ACM Conference on International Computing Education Research-Volume 2* (pp. 24–25).

Broll, B., Lédeczi, A., Volgyesi, P., Sallai, J., Maroti, M., Carrillo, A., & Lu, M. (2017). A visual programming environment for learning distributed programming. In *Proceedings of the 2017 ACM SIGCSE technical symposium on computer science education* (pp. 81–86).

Cannell, C., Tofel-Grehl, C., & Searle, K. (2020). Using Circuit Playground Express and Maps To Visualize Population Migration Data. In *Proceedings of ICLS 2020*.

Chapman, S., & Vivian, R. (2017). *Engaging the future of STEM: A study of international best practice for promoting the participation of young people, particularly girls, in science, technology, engineering and maths (STEM)*. Chief Executive Women.

Ceci, S. J., Ginther, D. K., Kahn, S., & Williams, W. M. (2014). Women in academic science: A changing landscape. *Psychological Science in the Public Interest, 15*(3), 75–141.

Code.org, CSTA, & ECEP Alliance (2022). 2022 State of Computer Science Education: Understanding Our National Imperative. https://advocacy.code.org/stateofcs

Cordero, E. C., Centeno, D., & Todd, A. M. (2020). The role of climate change education on individual lifetime carbon emissions. *PloS One, 15*(2), e0206266.

Davis, K., White, S., Madkins, T. C., & Sobomehin, O. (2021, March). Reimagining equitable computer science education: Culturally relevant computing in practice. In *Proceedings of the 52nd ACM Technical Symposium on Computer Science Education* (pp. 608–609).

Dijkstra, E. W. (1982). *Selected writings on computing: A personal perspective.* Springer-Verlag.

Engle, R. A., Lam, D. P., Meyer, X. S., & Nix, S. E. (2012). How does expansive framing promote transfer? Several proposed explanations and a research agenda for investigating them. *Educational Psychologist, 47*(3), 215–231.

Erickson, T., Wilkerson, M. H., Finzer, W., & Reichsman, F. (2019). Data moves. *Technology Innovations in Statistics Education, 12*(1). https://doi.org/10.5070/T5121038001

Ericson, B. J., Denny, P., Prather, J., Duran, R., Hellas, A., Leinonen, J., & Rodger, S. H. (2022). Parsons problems and beyond: systematic literature review and empirical study designs. *Proceedings of the 2022 working group reports on innovation and technology in computer science education* (pp. 191–234).

Estrada, M., Burnett, M., Campbell, A. G., Campbell, P. B., Denetclaw, W. F., Gutiérrez, C. G., & Zavala, M. (2016). Improving underrepresented minority student persistence in STEM. *CBE—Life Sciences Education, 15*(3), es5.

Fagliano, J. A., & Diez Roux, A. V. (2018). Climate change, urban health, and the promotion of health equity. *PLoS Medicine, 15*(7), e1002621.

Fields, D., & Kafai, Y. (2020). Hard fun with hands-on constructionist project-based learning. In S. Grover (Ed.), *Computer science in K-12: An A to Z handbook on teaching programming.* Edfinity.

Fields, D., Vasudevan, V., & Kafai, Y. B. (2015). The programmers' collective: Fostering participatory culture by making music videos in a high school Scratch coding workshop. *Interactive Learning Environments, 23*(5), 613–633.

Fields, D. A., Quirke, L., Amely, J., & Maughan, J. (2016). Combining big data and thick data analyses for understanding youth learning trajectories in a summer coding camp. In *Proceedings of the 47th ACM technical symposium on computing science education* (pp. 150–155).

Finzer, W. (2013). The data science education dilemma. *Technology Innovations in Statistics Education, 7*(2). https://escholarship.org/uc/item/7gv0q9dc

Finzer, W. (2016). Common online data analysis platform (CODAP). *The Concord Consortium.* www.concord.org/codap

Fisher, A., & Margolis, J. (2003). Unlocking the clubhouse: Women in computing. In *Proceedings of the 34th SIGCSE technical symposium on computer science education* (p. 23).

Goldenberg, P., Mark, J., Harvey, B., Cuoco, A., & Fries, M. (2020). Design Principles behind Beauty and Joy of Computing. In *Proceedings of the 51st ACM technical symposium on computer science education* (pp. 220–226).

Google LLC and Gallup Inc. (2020). *Current perspectives and continuing challenges in computer science education in US K-12 schools.* https://services.google.com/fh/files/misc/computer-science-education-in-us-k12schools-2020-report.pdf

Gransbury, I., Root, E., Brock, J., Cateté, V., Barnes, T., Grover, S., & Ledeczi, A. (2023, in submission). *A modular software engineering curriculum for secondary students.* Submitted to the 2023 international conference on software engineering (ICSE 2023).

Grover, S., Biswas, G., Dickes, A., Farris, A., Sengupta, P., Covitt, B., Gunckel, K., Berkowitz, A., Moore, J., Irgens, G. A., Horn, M., Wilensky, U., Metcalf, S., Jeon, S., Dede, C., Puttick, G., Bernstein, D., Wendell, K., Danahy, E., Cassidy, M., Shaw, F., Damelin, D., Roderick, S., Stephens, A. L., Shin, N., Lee, I., Anderson, E., Dominguez, X., Vahey, P., Yadav, A., Rich, K., Schwarz, C., Larimore, R., & Blikstein, P. (2020). *Integrating STEM and Computing in PK-12: Operationalizing Computational Thinking for STEM Learning and Assessment.* In The Interdisciplinarity of the Learning Sciences, 14th International Conference of the Learning Sciences (ICLS) 2020, 3, 1479–1486. Retrieved from https://par.nsf.gov/biblio/10289291

Grover, S. (2021a). Computational thinking today. In *Computational thinking in education* (pp. 18–40). Routledge.

Grover, S. (2021b). *"CTIntegration": A conceptual framework guiding design and analysis of integration of computing and computational thinking into school subjects.* https://edarxiv.org/eg8n5/.

Grover, S. (2021c). Teaching and assessing for transfer from block-to-text programming in middle school computer science. In *Transfer of learning* (pp. 251–276). Springer.

Grover, S., Oster, J., Ledeczi, A., Broll, B., & Deweese, M. (2022). Climate science, data science and distributed computing to build teen students' positive perceptions of CS. In Proceedings of the 53rd ACM technical symposium on computer science education V. 2 (pp. 1101–1101).

Hestness, E., McDonald, R. C., Breslyn, W., McGinnis, J. R., & Mouza, C. (2014). Science teacher professional development in climate change education informed by the next generation science standards. *Journal of Geoscience Education, 62*(3), 319–329.

Israel, M., Jeong, G., Ray, M., & Lash, T. (2020). Teaching elementary computer science through universal design for learning. In *Proceedings of the 51st ACM technical symposium on computer science education* (pp. 1220–1226).

Jean, D., Grover, S., Ledeczi, A., & Broll, B. (2023). MobileIoT for teaching "Internet of Things': Smartphones to promote accessible, engaging, and authentic experiences. In *Proceedings of the ISLS annual meeting, 2023 (ICLS)*. ISLS.

Kastens, K., Krumhansl, R., & Baker, I. (2015). Thinking big. *The Science Teacher, 82*(5), 25.

Kirk, C. M., Lewis, R. K., Brown, K., Nilsen, C., & Colvin, D. Q. (2012). The gender gap in educational expectations among youth in the foster care system. *Children and Youth Services Review, 34*(9), 1683–1688.

Khan, N. Z., & Luxton-Reilly, A. (2016). Is computing for social good the solution to closing the gender gap in computer science?. In *Proceedings of the Australasian computer science week multiconference* (pp. 1–5).

Krishnamurthi, S., & Fisler, K. (2020). Data-centricity: A challenge and opportunity for computing education. *Communications of the ACM, 63*(8), 24–26.

Ladson-Billings, G. (2014). Culturally relevant pedagogy 2.0: Aka the remix. *Harvard Educational Review, 84*(1), 74–84.

Lee, O. (2020). Making everyday phenomena phenomenal: Using phenomena to promote equity in science instruction. *Science and Children, 58*(1), 56–61. https://www.nsta.org/science-and-children/science-and-children-septemberoctober-2020/making-everyday-phenomena

Lee, O., & Grapin, S. E. (2022). The role of phenomena and problems in science and STEM education: Traditional, contemporary, and future approaches. *Journal of Research in Science Teaching, 59*(7), 1301–1309.

Lee, V. R. (2019). On researching activity tracking to support learning: a retrospective. *Information and Learning Sciences*, (120(1/2), 133–154.

Lee, V. R. (2018). On researching activity tracking to support learning: A retrospective. *Information and Learning Sciences*, 133–154.

Lee, V. R., Wilkerson, M. H., & Lanouette, K. (2021). A call for a humanistic stance toward K–12 data science education. *Educational Researcher, 50*(9), 664–672.

Lewis, C. M., & Shah, N. (2015). How equity and inequity can emerge in pair programming. In *Proceedings of the eleventh annual international conference on international computing education research* (pp. 41–50).

Limke, A., Milliken, A., Cateté, V., Gransbury, I., Isvik, A., Price, T., & Barnes, T. (2022). Case studies on the use of storyboarding by Novice Programmers. In *Proceedings of the 27th ACM conference on innovation and technology in computer science education* Vol. 1, pp. 318–324.

Litts, B. K., Searle, K. A., Brayboy, B. M., & Kafai, Y. B. (2021). Computing for all?: Examining critical biases in computational tools for learning. *British Journal of Educational Technology, 52*(2), 842–857.

Loudenback, T., & Jackson, A. (2017). The 10 most critical problems in the world, according to millennials. https://www.businessinsider.com/world-economic-forum-world-biggest-problems-concerning-millennials-2016-8

Lytle, N., Catete, V., Isvik, A., Boulden, D., Dong, Y., Wiebe, E., & Barnes, T. (2019a). From "Use" to "Choose" scaffolding CT curricula and exploring student choices while programming (practical report). In *Proceedings of the 14th workshop in primary and secondary computing education* (pp. 1–6).

Lytle, N., Dong, Y., Cateté, V., Milliken, A., Isvik, A., & Barnes, T. (2019b). Position: Scaffolded coding activities afforded by block-based environments. In *Proceedings of the 2019 IEEE blocks and beyond workshop (B&B)* (pp. 5–7).

Lytle, N., Milliken, A., Cateté, V., & Barnes, T. (2020). Investigating different assignment designs to promote collaboration in block-based environments. In *Proceedings of the 51st ACM technical symposium on computer science education* (pp. 832–838).

Madkins, T. C., Thomas, J. O., Solyom, J., Goode, J., & McAlear, F. (2020). Learner-centered and culturally relevant pedagogy. *Computer science in K-12: An A-to-Z handbook on teaching programming*, 125–129.

Mahmoud, Q. H. (2005). Revitalizing computing science education. *Computer, 38*(5), 100–99.

Margolis, J., Estrella, R., Goode, J., Holme, J. J., & Nao, K. (2008). "Claimed spaces: "Preparatory privilege" and high school computer science". In *Stuck in the shallow end: Education, race, and computing* (pp. 71–95). MIT Press.

Margolis, J., & Fisher, A. (2002). *Unlocking the clubhouse: Women in computing*. MIT Press.

Morales-Chicas, J., Castillo, M., Bernal, I., Ramos, P., & Guzman, B. L. (2019). Computing with relevance and purpose: A review of culturally relevant education in computing. *International Journal of Multicultural Education, 21*(1), 125–155.

National Science Foundation (NSF, 2017). *Big Data @ NSF*. https://www.nsf.gov/cise/bigdata/.

NGSS Lead States. (2013). *Next generation science standards: For states, by states (Vol. 1) Washington*.

Papastergiou, M. (2008). Are computer science and information technology still masculine fields? High school students' perceptions and career choices. *Computers and Education, 51*(2), 594–608.

Parsons, D., & Haden, P. (2006). Parson's programming puzzles: A fun and effective learning tool for first programming courses. In *Proceedings of the 8th Australasian conference on CS Ed-Volume 52* (pp. 157–163).

Pollock, L., McCoy, K., Carberry, S., Hundigopal, N., & You, X. (2004). Increasing high school girls' self confidence and awareness of CS through a positive summer experience. *ACM SIGCSE Bulletin, 36*(1), 185–189.

Register, Y., & Ko, A. J. (2020, August). Learning machine learning with personal data helps stakeholders ground advocacy arguments in model mechanics. In *Proceedings of the 2020 ACM Conference on International Computing Education Research* (pp. 67–78).

Repenning, A., Webb, D. C., Koh, K. H., Nickerson, H., Miller, S. B., Brand, C., Horses, I. H. M., Basawapatna, A., Gluck, F., Grover, R., & Gutierrez, K. (2015). Scalable game design: A strategy to bring systemic computer science education to schools through game design and simulation creation. *ACM Transactions on Computing Education (TOCE), 15*(2), 1–31.

Resnick, M., Berg, R., & Eisenberg, M. (2000). Beyond black boxes: Bringing transparency and aesthetics back to scientific investigation. *The Journal of the Learning Sciences, 9*(1), 7–30.

Resnick, M., & Silverman, B., (2005). Some reflections on designing construction kits for kids. In *Proceedings of the 2005 conference on interaction design and children*, 117–122.

Rosenberg, J. M., Lawson, M., Anderson, D. J., Jones, R. S., & Rutherford, T. (2020). Making data science count in and for education. In *Research methods in learning design and technology* (pp. 94–110). Routledge.

Sentance, S., Waite, J., Yeomans, L., & MacLeod, E. (2017, November). Teaching with physical computing devices: the BBC micro: bit initiative. In *Proceedings of the 12th Workshop on Primary and Secondary Computing Education* (pp. 87–96).

Smith, M. (2016). *Computer Science for All [blog]*. https://obam-awhitehouse.archives. gov/blog/2016/01/30/computer-science-all

*Snap!: A visual, drag-and-drop programming language.* http://snap.berkeley.edu/snapsource/snap.html.

Srikant, S., & Aggarwal, V. (2017). Introducing data science to school kids. In *Proceedings of the 2017 ACM SIGCSE technical symposium on computer science education* (pp. 561–566).

Sun, C., & Clarke-Midura, J. (2022). Recruiting K-12 youth into computer science: Insights from a multi-year research project. *ACM Inroads, 13*(2), 22–29.

Touretzky, D., Gardner-McCune, C., Martin, F., & Seehorn, D. (2019). Envisioning AI for K-12: What should every child know about AI?. *Proceedings of the AAAI conference on Artificial Intelligence, 33*(01), 9795–9799. https://doi.org/10.1609/aaai. v33i01.33019795

Weintrop, D., Killen, H., & Franke, B. E. (2018). *Blocks or text? How programming language modality makes a difference in assessing underrepresented populations*. International Society of the Learning Sciences, Inc. [ISLS].

Weintrop, D., Killen, H., Munzar, T., & Franke, B. (2019, February). Block-based comprehension: Exploring and explaining student outcomes from a read-only block-based exam. In *Proceedings of the 50th ACM technical symposium on Computer Science Education* (pp. 1218–1224).

Zhi, R., Chi, M., Barnes, T., & Price, T. W. (2019). Evaluating the effectiveness of parsons problems for block-based programming. In *Proceedings of the 2019 ACM conference on international computing education research* (pp. 51–59).

# 5

# THE CASE FOR COMMUNITY-CENTERED DATA SCIENCE

*Colby Tofel-Grehl, Tyler Hansen, Emily Slater, and David Feldon*

Data drives policymaking and our modern economy. It is imbued with substantial societal power, but the ability to access and steer that power is not equitably distributed. Not only do constructed metrics and algorithms frequently perpetuate biases in ways that can sustain inequity, but the education and empowerment necessary to prepare students to recognize and correct these injustices are often withheld from the communities who need them most. Efforts to bring data science education to historically marginalized communities hold promise to redistribute access to the necessary funds of knowledge (D'ignazio and Klein, 2023). However, as is often the case with STEM (science, technology, engineering, mathematics) disciplines, relevant knowledge can be too easily presented in ways that exclude the values, identities, and cultures of intended recipients. STEM is frequently viewed as a space belonging to the white, straight, male, and affluent, into which others may be invited but remain "guests" who are not afforded the right to shape the dominant epistemology to include their own ways of knowing (Bang and Medin, 2010; Calabrese Barton and Tan, 2019; Carlone and Johnson, 2007). Accordingly, educational researchers have increasingly pursued the development of culturally responsive curriculum to position STEM learning in ways that are not exclusionary of the identities of the learners (Ladson-Billings, 2021). However, such efforts frequently are led and resourced by individuals and institutions who come to marginalized communities as outsiders (Bridges and Bridges, 2017) without the cultural understanding to optimize or evaluate the responsiveness of resulting educational materials and opportunities. Even when codesign practices are engaged, power imbalances persist. Researchers bring with them the authority of advanced degrees and incentivizing resources that are often desperately needed by stakeholders within

DOI: 10.4324/9781003364634-5

communities (Kyza and Agesilaou, 2022). Further, culture itself holds many tacit understandings, entailing meanings, signs, and tools that have evolved to meet the specific needs of people with a shared history that cannot be easily or fully conveyed to newcomers (Ladson-Billings, 1995).

## Positionality Statement

As white researchers working with Indigenous Hawaiian youth, we have spent a great deal of time considering what it means to be culturally responsive. In our discussions and reflections, we find ourselves in a catch-22 when it comes to concepts such as "culturally responsive" and "culturally affirming". Despite our goal to be culturally responsive, we recognize ourselves to be outside the culture with which we work. Furthermore, Hawaiian culture and community offer many layers of nuance within and across island spaces. Our outsider status affords us both privileges and challenges. Being outside the culture, we retain an outsider's perspective and apply understandings that are rooted in different perspectives. However, it also limits our ability to speak to the nuanced impacts of specific educational practices on members of said culture. We sat in the discomfort this disconnect created and came to believe that for us to speak of the cultural responsiveness of our work with Indigenous youth created the potential to affirm whiteness in an unintentional way. As Hawkman (2020, p. 404) explains:

> Whiteness is an ever-shifting, hierarchical, hegemonic power structure and identity construct that informs the ways individuals view themselves and society ... Whiteness has also been described as the water in which white people swim (Owen, 2007). It is all around them, keeping them afloat.

While it is important to reflect on our work and ensure that it did not create barriers or scaffold systemic disenfranchisement, we also did not feel comfortable evaluating how our work affirmed another group's culture; to do so risks co-opting or misinterpreting the words and perspectives of another culture's values. Further, we recognize the extent to which individual students' affinities with the culture of their community vary due to personal, familial, geographic, and socioeconomic factors (Moscovici, 1998). Accordingly, we wanted to avoid both essentializing the relationship and understanding that students may have with their shared culture and relying on a presumption that all students participating in a lesson might resonate with "culturally responsive" content in similar ways.

For these reasons, we conceptualize our work as centering community rather than being responsive to culture. Having a long relationship with the specific community partnering in the present study, we felt that centering community aligned best with both our goals and our recognized limitations in understanding and supporting its nuanced and multifaceted culture. Additionally, community centering may be more concrete in addressing local needs than cultural responsiveness. It inherently connects across the three layers of Lee, Wilkerson, and Lanouette's (2021) humanistic stance toward data science education ((i.e., personal data layer, cultural data layer, and sociopolitical layer). It also emphasizes cultural funds of knowledge as they manifest contemporaneously for youth rather than in a form more reflective of a potentially distal historic legacy.

While many members of the community identified as Indigenous, others held different historically and culturally distinct identities, which highlighted the challenge of being culturally responsive while ensuring full inclusion for students from starkly different marginalized groups. However, all students identified as sharing community, which they conceptualized as (in the words of one student) "all the people who help each other here". From this framing, which reflects the core value of *kōkua* (spirit of kindness accompanied by a desire to help one another, without expecting anything in return) from Hawaiian culture, we focused our curriculum on the concerns and needs of the community toward accomplishing their goals. This understanding permits us to engage in developing community-focused and supportive curriculum without requiring us, as outsiders, to frame or evaluate the cultural responsiveness of its content. The case study in this chapter contributes to principles of development and enactment needed by the field of data science education. In it, we identify central pillars necessary for engaging in community-centered data science education and frame guiding questions that can support intentional curricular design supportive of student and community meaning making around data science.

Working with middle school students across grades 6–8 in rural Hawaii, we engaged in codesign to develop a curricular unit on invasive species that centered data science practices, as laid out in the NGSS. Through codesigning with local educators, we learned about *kilo*, a set of Hawaiian community practices that center nature, observation, and meaning making in support and stewardship of the land. The curricular unit was deployed in the summer of 2023 within a local STEM and computing (STEM+C) summer program. The class ran for 10 days and included naturalistic observations and data collection local to the school and home setting of the community of focus.

## Intersecting Frameworks

We frame this work through a power-conscious synthesis of rightful presence within STEM as described by Calabrese Barton and Tan (2019, 2020) and Lee et al.'s (2021) humanistic framework for data science. Broadly speaking, data science can be thought of as the application of mathematical, statistical, and reasoning skills to extract meaning from data. Within communal settings such as schools, it is essential to understand who is constructing that meaning and what perceptions, experiences, and realities influence their process. Thus, we engage Calabrese Barton and Tan's conception of rightful presence to center the experiences and understandings of community members around data science within this chapter.

### Rightful Presence

Rightfully present school spaces center questions and concerns of the youth in those spaces, acknowledging "legitimate membership in a classroom community because of who one is (not who one should be), in which the practices of that community work toward and support restructuring power dynamics toward more just ends through making injustice and social change visible" (Calabrese Barton and Tan, 2019, p. 618). By centering the rights, experiences, and realities of youth, the rightfully present space allows the varied and nuanced identities youth embody to be entirely and jointly present in the school space, "making [them] visible and present in teaching and learning alongside the amplifications of youths' lives and wisdom" (Calabrese Barton and Tan, 2020, p. 436). Rightful presence also makes the inherent political struggles that communities experience present and central within schooling. Rightfully present data science learning requires the acknowledgement of and engagement with the political nature of the endeavor. While many espouse the notion that data and data science are objective, rightfully present data science recognizes and articulates the marriage between the political and the scientific that can bias understandings and privilege some groups at the expense of others.

### Humanistic Framework for Data Science

Lee et al. (2021) identify three layers of scholarly work across fields that engage data science work centering youth and youth learning: the personal data layer, the cultural data layer, and the sociopolitical layer. Personal data can be defined as data that a student directly engages. Examples include students collecting data or directly evaluating or manipulating data to understand a larger concept. The cultural layer examines the tools and practices typically used by scientists and other data practitioners, as well as how cultural norms might influence the ways in which data are conveyed and interpreted. A classic example is the scientific tendency to reify evidence as "real" independent of the perspective of any observer (Manz, 2016). Finally, the sociopolitical layer involves bringing a critical lens into how data is created, filtered, and understood through structures of power.

The personal layer of data can position students as holding foundational agency in the construction and analysis of datasets. When data science opportunities are offered in a manner that fosters rightful presence, students can collect and represent data anchored in the needs of their communities that they articulate, making decisions using their own community knowledge to make informed decisions about where, when, and how to collect data that they will find meaningful and useful. Similarly, the cultural layer provides the opportunity for students to make sense of the data in ways most responsive to the needs and issues of their communities while shaping the representation and communication of those findings in ways that can offer maximum impact for community stakeholders. In both layers, students can put forward and leverage the knowledge and insight of their own experiences to produce knowledge that is locally meaningful and empowers them to author their own legitimacy as knowledge producers. As students engage the sociopolitical layer, they extend the rightfulness within which they engage data. Through the intentional interrogation of the frameworks of sensemaking and the structure of the data tools they adapt to their own purposes, students can reconfigure frameworks to reflect their understandings and values, disrupting "normative knowledge/power relationalities" (Calabrese Barton and Tan, 2020, p. 437) that often constrain the ability of marginalized communities to engage assumptions that match their lived realities.

The personal data layer positions students as more than passive observers, encouraging them to become active contributors who collect and visualize data that aligns with their community's priorities. By infusing personal agency into data construction and analysis, students forge a stronger connection between data and community needs. Moving to the cultural data layer, a gateway emerges between technical mastery and cultural understanding. This layer equips students with the ability to interpret data and communicate findings in ways that resonate with their community's distinct challenges. This approach ensures data's transformation, promoting community responsiveness and empowerment. The sociopolitical layer acts as a fulcrum for redefining power dynamics. This layer describes empowering communities to redefine their own narratives, fostering an environment where data becomes a conduit for equitable representation and social justice. These layers converge to train students to become adept at integrating data into the dynamics of their communities.

### Community-Centered Curriculum

In shaping a data science education approach, temporal and spatial dimensions emerge as important considerations. Community-centered education serves as a perspective attuned to the dynamic nature of culture as it is practiced within a time and space. Diverging from culturally-responsive strategies, which can operate within more static cultural contexts, a community-centered approach includes specific temporal moments and distinctive spatial contexts through which youth experience and understand their community. We sat in the discomfort this disconnect created. We came to believe it inappropriate for us to speak of the cultural responsiveness of our own work. To do so could unintentionally affirm whiteness and white perspectives.

To center community, we operationalize key elements based on the distinction between cultural settings and cultural models. Gallimore and Goldenberg (2001, p. 48) explain that "culture exists (and is created) in settings, those occasions where people come together to carry out joint activity that accomplishes something they value". These settings serve as a dynamic model for how culture is practiced within a time and space. They provide a unit to evaluate interactions within contexts where culture connects with learning experiences. In parallel, cultural models are the belief structures that frame individuals' perceptions, behaviors, and interactions within their learning landscapes through the lens of their culture. By arguing that community is a type of cultural setting, we eschew excluding any member of the community on grounds of cultural affinity or representativeness. In doing so, we hone our curriculum development and analysis to community values and needs.

Through engagement with rural Hawaiian youth in data science practices to make sense of their evolving local ecosystem, we sought to answer the following research questions:

1　How do the features of the developed data science curriculum manifest aspects of community-centeredness?
2　How do students experience community-centered data science through the curriculum?

## Methods

To understand the approach to this community-centered work, we present rich detail about its context.

Community of Focus. The community of focus centered in this work is a rural town, Nui Huna, located on the big island of Hawaii. There is a particularly salient dynamic between the community of focus and the broader global science community. Nestled in the foothills of the Mauna Kea, the town of Nui Huna has historically been a hub of ranching and farm life on the island. Recently, the town has also been host to several significant protests over the building of a new telescope on the Mauna Kea. While the Mauna Kea is situated to be a great resource to the field of astronomy, it is also the holiest site to Indigenous Hawaiian religious practitioners. The building of the telescope is considered a desecration of sacred lands and has been the source of great animus between local community members and scientists. Youth at the community school refer to science as *pau*, Hawaiian slang for "done" (Tofel-Grehl, 2023).

The school. Nui Huna Middle School (NHMS) is the lone public middle school in the town. Serving roughly 200 students in grades 6–8, NHMS is a rurally located Title 1 school that has experienced significant fiscal hardships since the COVID 19 pandemic. Due to the pandemic, school enrollment has been steadily in decline over the past three years as families moved from the rural space to urban ones for economic reasons. This decline in enrollment causes deep school community challenges as per-pupil dollar allocations from the state do not easily cover the needed faculty and staff costs across those three grades. However, as the only local public educational opportunity for youth in the area, the need for the school is great. Adding an additional layer, the school's core mission focuses on supporting and fostering Hawaiian culture, community, and language within the school. These factors make the school an essential resource and hub for the local community.

Summer program. The summer school program is part of an 8-year research practice partnership (RPP) between the NHMS and the lead author. The focus of this research practice is to scaffold and support teacher professional learning in STEM+C integration. The program currently serves approximately 45 students each summer and engaged approximately 15 teachers in scaffolded professional development. Drafts of curricular units and classes are developed by the research team in partnership with the participating teachers. Teachers then modify and develop the materials to suit the needs of their students for further deployment during the school year.

Participants. Forty-five students enrolled in the summer program in 2023. Of those students, 29 consented to participate in the research. Seven students were selected for follow up interviews about their experiences and perspectives on the summer program.

Coqui in Hawaii. Coqui frogs are an invasive species in the area of Nui Huna. Many adults see them as a nuisance that lowers property value due to noise pollution. Within youths' reflection on the Coqui, they were welcomed to explore the feelings they hold about the frogs compared to those they hear across their community from adults.

Data sources. Data sources for this analysis included student interviews, classroom artifacts, observational field notes, and student collected data around coqui.

## Curriculum

The Coqui Frog Curriculum was part of a summer school program developed in partnership with the target community and school. This was the third year of deployment of the summer program that sits at the center of the RPP at the heart of this research. We built the curriculum to engage students in geographically and community relevant data science practice. Specifically, we focused

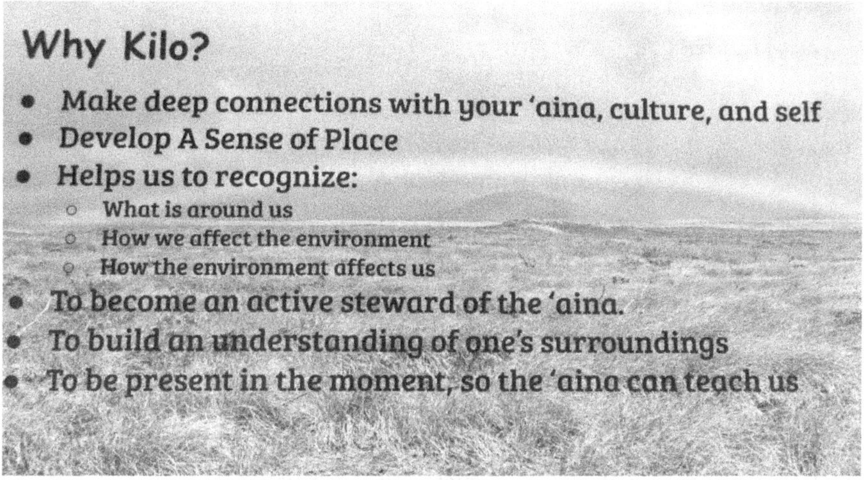

## Why Kilo?

- **Make deep connections with your 'aina, culture, and self**
- **Develop A Sense of Place**
- **Helps us to recognize:**
  - What is around us
  - How we affect the environment
  - How the environment affects us
- **To become an active steward of the 'aina.**
- **To build an understanding of one's surroundings**
- **To be present in the moment, so the 'aina can teach us**

FIGURE 5.1    Example of participating teacher's Kilo materials.

on the process of collecting data about the natural world around them. When we provided the draft concept materials to the master teacher responsible for deploying the curriculum, he mentioned that another one of our master teachers had developed a science lesson centering the Hawaiian community practice of observation called "*kilo*". When he showed us her materials, he said "It literally says *kilo* is data!" Another teacher participating in the program showed us the materials that she had made (see Figure 5.1). When asked about what she had already taught the students about *kilo*, she stated that "We have been practicing it here at school, starting from our [name of workshop] like 15 years ago. It's been a working document between the [other schools] and myself". Leveraging this existing practice and lesson, we integrated *kilo* into our youth led data collection process.

During curricular development, we were able to identify four key connections between *kilo* and standard data science practices. These connections made *kilo* a part of our community-centered data collection process. In Hawaiian, *kilo* can be defined as to observe or forecast. It can be used to describe the process of observation or describe a person who is particularly adept at those skills. However, it is also considered to be the process of using keen observational skills used to observe the natural world and make decisions; this practice has existed for generations within Hawaiian communities (Morishige et al., 2018). *Kilo* can be used to describe the practice of making observations or to refer to the person making the observations themselves. For example, a *kilo* (observer) might stay on shore to signal the direction that schools of fish were moving to fishermen on a reef (Titcomb, 1972). One historically salient example of *kilo*

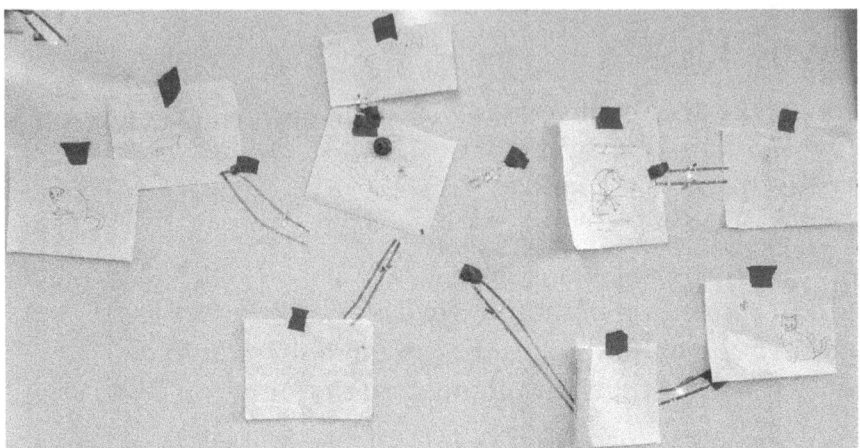

**FIGURE 5.2**   Example of ecological interactions with circuits.

within Hawaiian history would be *Kilo Hōkū*, or Hawaiian Wayfinding by ce-lestial observation. This practice entails open-water navigation without using instruments (i.e., using constellations to navigate the ocean).

The summer program offered students four discrete classes that created STEM+C learning experiences. Within the Coqui Frog class, students first explored the ecological interactions of the coqui introduction by creating an electronic food web. Students selected an animal of their choosing and read about the nuanced interactions that have been documented or hypothesized. They used LED lights and conductive thread to visualize the relationship. As shown in Figure 5.2, students were able to choose the description of each color of LED.

From there, students were introduced to writing down their *kilo* experiences. While outside, they were encouraged to notice the environment around them and notice their thoughts and feelings in relation to the outside environment (see Figure 5.3).

Additionally, they measured temperature and humidity using inexpensive thermometers and hygrometers. They recorded the information in their field jour-nals. After collecting these covariates, students were ready to collect population data on the coqui. Their homework was to practice *kilo* outside at night when the coqui are most active and calling. They documented their experiences as data and audio-recorded the coqui calls for around 10 minutes. These activities ap-plied Lee and colleagues' (2021) first layer, as students not only collected data, but create the variables of data that they collect. The following day, students used a teachable machine programmed to detect coqui calls to determine population density. Students were able to choose which covariates would be used to explain

**FIGURE 5.3**  Example of *kilo* journal entry.

the coqui density, along with using time of day, temperature, and humidity (see Figure 5.4).

Students then used a combination of pictures and codable circuits to visualize their findings on a printed map of their area. By using programmable circuits, students conveyed information in novel ways, such as making lights blink in coqui hotspots (see Figure 5.5).

Students then used this visualization to make inferences about the coqui density and their covariate data. These activities engaged the cultural layer of Lee and colleagues' (2021) data practices, as students both practiced being data scientists and articulated the types of data engaged. Additionally, the political layer was also explicitly laid out at both the beginning and end of the curriculum when students were asked to reflect on their feelings about the classic coqui call. Interestingly, students had mixed feelings about the coqui noise, with most students listing both good and bad feelings about it. We decided to show students the dichotomy between how the Hawaiians typically perceive the coqui with the perceptions of Puerto Rican peoples. In Puerto Rico, coqui are portrayed as a source of local

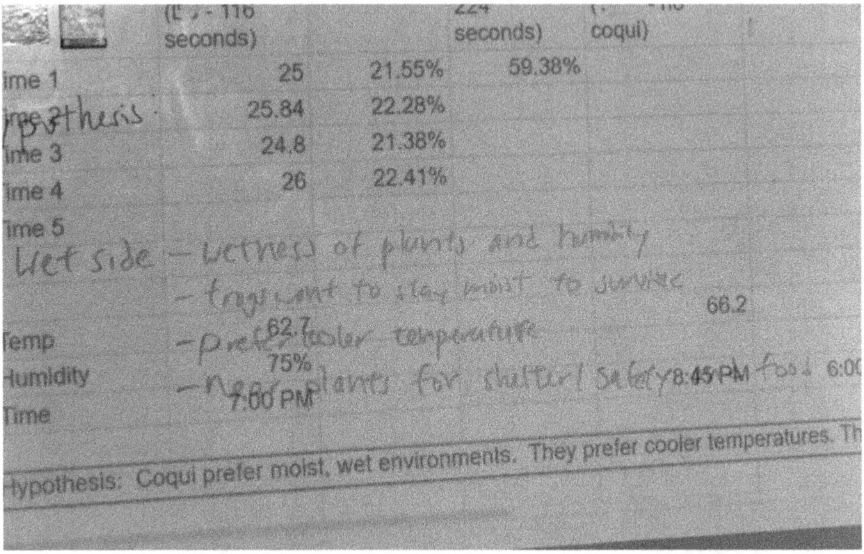

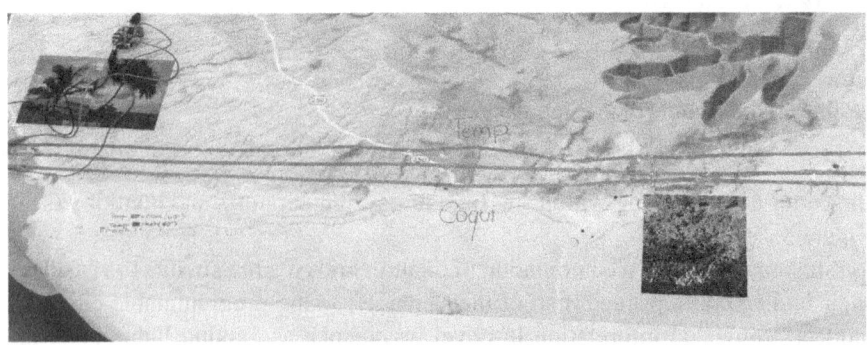

| | ($\mathrel{E} \mathrel{\cdot} 116$ seconds) | | | seconds) | coqui) | |
|---|---|---|---|---|---|---|
| ime 1 | 25 | 21.55% | 59.38% | | | |
| ime 3 | 25.84 | 22.28% | | | | |
| ime 3 | 24.8 | 21.38% | | | | |
| ime 4 | 26 | 22.41% | | | | |
| ime 5 | | | | | | |

*Wet side – wetness of plants and humidity*
*– fragiment to stay moist to survive*
*– prefer cooler temperature*                    66.2
*– near plants for shelter / safety food*

| Temp | 62.7 |
| Humidity | 75% |
| Time | 7:00 PM    8:45 PM    6:00 |

*Hypothesis: Coqui prefer moist, wet environments. They prefer cooler temperatures. Th*

**FIGURE 5.4**  Group of students working through data analysis and covariate identification.

**FIGURE 5.5**  Student built data visualization of coqui population density.

pride, whereas Hawaiian's think of them as pests. Through reflection about environmental stewardship, they considered the implications of their perspectives.

## Analysis/Findings

In this case analysis, we first delineate the community-centered data science practices within *kilo*. These practices make up the heart of this case study as the connections made within and across practices informed the larger

TABLE 5.1 Connecting kilo to standardized science and data science practices

| Kilo Practice | Science and Data Science Practice |
| --- | --- |
| Making observations | Making observations; field journaling |
| Forecasting and predicting | Making predictions |
| Fostering a relationship with nature | Journaling, making observations, and determining covariates from observation |
| Stewardship | Interpretation and fostering of rightful presence |

pillars set forth in the second part of this analysis. From the connections made between *kilo* specific practices and larger more generalized data science practices, we inductively established the pillars for our future work on community-centered data science learning and the associated questions we could use as reflective tools for informing our work. Table 5.1 maps out the connections between *kilo* practices and the science and data science practices laid out in the NGSS.

### Connected Practice 1. Observations

As noted earlier, we identified four facets of *Kilo* connected to standard data science practices. First, central to *kilo* broadly is making observations. *Kilo* means observing the environment as part of a specific community. Within our lessons on coqui frogs, we connected the practice of making observations to keeping a field journal. Our youth scientists engaged in the *kilo* practice of observing nature and used their field journals to collect covariate data such as temperature and humidity (see Figure 5.3 of student's *kilo* journal). We intentionally connected the mindfulness practices laid out in *kilo* to youth's process of collecting covariate data.

### Connected Practice 2. Predictions

*Kilo* also means to forecast. This is done as an intentional practice to understand the environment around and to predict natural phenomena. This meaning of *kilo* connected strongly with the standardized practices around data interpretation. Seasons and weather patterns are typical phenomena of focus for *kilo* practitioners. Within our community of focus, we applied this process of meaning making to the interpretation of data collected. Coming to a conclusion based on data is a characteristic of both data analysis and *kilo*. As a historically oral tradition, there is also the connection between the personal data described by Lee and colleagues (2021) and the personal practices of *kilo*.

### Connected Practice 3. Connections

*Kilo* commands not only an observation of nature, but a relationship with it. Making subjective observations about oneself and the objective observations that come with it are a practice to foster that relationship. This also accurately describes personal data. It is data that is inherently meaningful to the individual. Field journal is an appropriate Western counterpart to this practice because it does not diminish the personal or cultural aspects of *kilo*. Rather, it encourages it.

### Connected Practice 4. Agency

Finally, stewardship is a large part of *kilo*, as one reflects on the relationships between people and the land around them. This data interpretation is both individual and community-wide. Reflecting on the relationship between humans and the land they steward requires interpreting information. Data practices only make these observations more explicit. "What is best for the land" is informed by data collection and deciding what is best brings in community or human considerations.

From the four connected practices identified between *kilo* and Western science and data science practices, we analyzed our data for central pillars of community-centered data science. Table 5.2 delineates these connections.

### Personal and Community

Student experiences with *kilo*. To understand how students made sense of the data science unit and *kilo* practices within their communities, we asked students to reflect on what constitutes "data", what *kilo* is, and if these practices relate to their everyday experiences and life in general. Some additional reflection questions asked students about the coqui population and invasive species.

Students clearly identified *kilo* as a practice that was both relevant to their personal lives and their community as it is practiced. For example, student B6 said:

That was cool because I usually just observe with my eyes and record it in my brain and be like "oh yeah, that's cool," and just forget about it eventually. But it was cool writing stuff down and drawing pictures because I really enjoy drawing. So that's my thing because my dad is an artist and my nana is an artist. So, that was cool getting to write it down in our little notebooks and taking specific time to just observe and not just like, observe while playing volleyball. I think it relates to my life as well because that kind of inspired me to do Kilo at home as well … I have a million notebooks, so I'm definitely

**TABLE 5.2** Connecting community-centered data science and layers of humanistic data science

| Pillars of Community-Centered Data Science | Curricular Enactment | Layer from Lee et al. | Example from Case Study |
|---|---|---|---|
| Center a specific community within contextual time and space. | Identification of community targeted, contextual situation, and curricular goals. | Cultural and sociopolitical layer | 1. Community of focus: Rural middle school students on the big island of Hawaii.<br>2. Context: Due to the conflict around the construction of a new 30M telescope on the Mauna Kea, science is not part of the community at this time (Tofel-Grehl, 2023).<br>3. Curricular goals: CS Standards and NGSS Standards HERE |
| Engage practices authentic to the community of focus. | Identify authentic community practices that align with data science practices or methods. | Personal and cultural | 1. Practices in Kilo that are present in Hawaiian culture: observation of the natural world.<br>2. Incorporating Kilo as the method of data collection. Encouraged by this process to identify things that are important to them and their surroundings. |
| Tackle problems or questions authentic to the community. | Open inquiry approach: Problem or question identification is done in partnership with the community. | Personal Cultural | 1. Personal: Youth experiences with coqui<br>2. Coqui is an invasive species on the big island of Hawaii.<br>3. Coqui invasion within that specific community.<br>4. Sociopolitical: Implications of coqui invasions evaluated. (Who is angry at the coqui population and why?) |
| Scaffold and foster the rightful presence of community. | Autonomy and authorship of decision making including variable selection, analysis decisions, and interpretations. | Personal | 1. Youth selection of plants as covariate.<br>2. Interpretation of data around the spaces coqui have taken over as well as the implications for property values and other community concerns.<br>3. Space provided to reflect on the thoughts and feelings toward coqui as an invasive species. |

going to start filling out. Um, some of them, most of them are empty. So I think, yeah, we did some Kilo, I did some *kilo* at home as well, and that was cool because I took 15 minutes off from doing anything else and just recorded in my notebook for a little while, and that was very peaceful and it kind of like helped me like observe my own house and my own yard better than I was doing before.

The elements of *kilo* and data collection practices were already part of her recreational schema. However, she clearly identified the merits of practicing Kilo as a means of understanding her surroundings and, subsequently, her community better than she had before. Further, the familiarity she had due to her family's artistry only furthered her desire to continue the community practice. Further, she understood the merits of *kilo* and data collection as an opportunity to notice the world around her, free from distractions. The sentiment of building a personal connection to the land around you was echoed by Student C8. He said:

Like you just sit out there listening to all the sounds and doing your *kilo*, just trying to see if we have any coqui. Like, it's just pretty cool to just sit out there and just listen. To [observe] what you see or hear … when I'm out by myself, I like the quiet and when I sit out there, like you could just hear everything. You hear the wind like flowing down at the beach, you hear the water, you hear people, you can basically hear pretty much everything.

Students built a connection between their own personal interests and experiences to the community. *Kilo* encouraged students to think about their existing knowledge about their natural surroundings in order to interpret the data that they were collecting. Although hygrometers and thermometers were provided to collect humidity and temperature data, many students had already hypothesized their results by way of their existing knowledge of their town. Student C7 stated that:

I'm pretty sure [there are many coqui] because my house is on the wet side; it's because of the moisture and the temperature, but also it isn't super cold there. I'm pretty sure coqui don't like to live in cold or super cold or super hot areas. So, [coqui] live in a slightly warm [environment].

### Problems in the Community

Students were quite adept at identifying invasive species as a local concern, even before instruction. They were able to not only define invasive species, but point to other invasive species and the specific detriment that they cause for the community. For instance, Student B6 said:

> An invasive species is an animal or a plant that's not native. And it's usually invasive species that are harming the native environment. So, like, there are a lot of invasive species in Hawaii…for example, the goats, which are destroying a lot of the plants on Mauna Kea. And it's really bad because there's so many of them.

Students largely repeated the same community concerns about the coqui invasions as adults, albeit with a few notable caveats. Student B7 described his concerns by saying "Where I live, [the coqui calls], it's nonstop. It's just chirp, chirp, chirp, chirp. It's not gonna stop. And never stops. So I have to cover my ears with my pillow". Student C3 repeated these concerns when noting "They do bother me though. Like, the sound that they make at night, I hear them all the time and it's so annoying". We can infer from these statements documenting student feelings about the coqui call that students thought critically about the impacts of the coqui invasion on the community in a nuanced fashion. Rather than simply condemning the coqui, students weighed the consequences of the invasion in terms of their personal experience and community. When asked what class was most relevant to them, Student D2 stated "Uh, the coqui frog class. It's animals and animals are part of our culture that we live here. Because since I was raised here, uh, they played a big role". By creating space for and engaging student concerns, our *Kilo* unit centered the temporal needs, considerations, and issues of our community of focus in ways that culturally responsive approaches may have missed.

### Rightful Presence

Students demonstrated rightful presence in a variety of ways. Students were clearly able to identify their own lived experience as a valuable asset to the investigation. This was evidenced by students making open predictions about the results of the investigation and immediate interpretations of the data they were collecting. As student C3 put it:

> My friend, [name redacted], she lives, uh, over on the wet side, so there'd be more coqui. So, we put, um, her house for red, which, um, which is, um, more coqui. And then for some people they had coqui, but it was fewer. So, we put yellow and green means zero coqui at all.

Many other students made similar comments, sharing their experiences as an integral part of data collection and interpretation. This was also evidenced by students' curiosity about their own chosen covariates. For example, when asked about *Kilo*, student A6, a female student in elementary school said:

> [*Kilo*] is, um, observing, like, I observed the 'Olena and this new leaf. It's like a circle. Like a little pipe. So I look inside it and there's a bunch of little fuzzy thingies inside and I wonder what it is. There's like a circle where the coqui frogs kind of stay away. And then there's like a larger circle where the coqui claim as their territory.

Students used their prior knowledge of local plant life learned within their community to make sense of coqui distribution. We not only infer that this had a positive shift in their autonomy, but in their identity as well. As Student B7 puts it:

> It's also pretty interesting because I'm learning new stuff and I'm not really a science-y person. When I was growing up, I didn't know how to do all this stuff, but like, but since I'm doing it right now, and if I wanted to be a scientist, I would already know all this stuff.

Students not only used their funds of knowledge in guiding data interpretation, but were also the arbiters of data visualization. They came up with their own methods of visualizing the distribution of the coqui population within their community. Specifically, students wanted to display "high" coqui populations as either red or green, and selected what those colors meant. Student B7 summarized their visualization "We circled where we live on the map and then we started putting lights down and I have a lot of coqui in my area and most people have like zero coqui at all cause they live on the dry side". Student B7 observed "Green lights mean that there's a lot of coqui, yellow means that there's a little bit of coqui, but not as much as green and red means that there's no coqui at all". Lastly, students were also able to make connections between invasive species and colonialism. Although the curriculum doesn't explicitly discuss invasive species as a direct consequence of Hawaiian annexation, students are clearly aware of it. As noted by student B6 "… there's a lot of invasive creatures in Hawaii because of all the people, the settlers who came and like brought all these diseases and these plants and these animals that weren't supposed to be here". Students had conflicting feelings about the coqui, but nearly all of them were able to identify the source of invasive species. By exploring their personal feelings, the source of invasive species, and learning about data, they were able to critically evaluate the sociopolitical implications of data science through the context of ecology.

Data Science and *Kilo* Comparisons. *Kilo* as a community practice easily translated into a method of data collection by students. When asked if students were working with data in their courses, Student B7 said:

> We did coding and stuff and we did like coqui sounds like recording coqui sounds, that's data. It's like we're tracking the noise. Like means something like, say like it means kind of like what this is. It means like what we would do for the coqui would map it.

This student clearly identified Lee et al.'s (2021) personal and cultural level of data. She recognized that she was collecting data, while also learning the data practices of science. Students clearly identified both *kilo* and data as a type of observation. Student C3 explicitly stated this:

> *Kilo* is, um, another type of data where you record it on a piece of paper or journal or whatever, and *kilo* pays attention to the surroundings, the area… *Kilo* is how you're feeling. *Kilo* three is paying attention to the trees, the plants, and *kilo* for the animals. So basically *kilo* kind of means paying attention to your surroundings. Um, I feel like it can connect to your life at school. Like, um, *kilo* is how you're feeling. [*Kilo*] can, uh, record that like in a journal or diary. Um, but I feel like it's something to, uh, I feel like it's a good thing to look, uh, to know how to do.

Students were also able to identify specific data science practices through *kilo*. This ranged from simply "[*Kilo*] kind of like, uh, intensifies my observation skills", to more specific aspects of data collection. For example, Student B6 reflected on her shift in her understanding of recording observations:

> I do a lot of art, so I observe things very carefully, but I usually just store it in my head and like, it's kind of cool that we're recording it in notebooks and then like, uh, taking pictures and uh, doing, uh, sound recordings of the coqui frogs. So I can draw what I see or write down like the description of that and like, I would do that occasionally, but usually I'm just like, oh yeah, that's cool. And then just store it out, weigh it in my brain and forget about it eventually.

In this way, we infer that she lifted the barrier between her artistic capabilities and data science practices through *kilo*. Students identified *kilo* as a community practice and incorporated their own capabilities into an authentic investigation of the coqui invasion.

## Conclusions

Our community-centered data science framework is grounded in the notion that culture interacts with learning experiences, such as with this codesigned data science curriculum. The cultures of science and data science engage people as insiders who have historically shaped the meaning, signs, and tools for practitioners in the present. The practices of science and data science within any given context depend on both the practices of people in the past and the understandings and uses of those currently engaged in the cultural activities. Specifically, science and data science as a cultural practice is dynamic and temporal. As we conducted our collaborative work, we sought ways to codesign curriculum with teachers that created data science activities about invasive species through shared values, community-centeredness, and human-centeredness.

The sharing of the *kilo* practices connected to our definition of science and data science practices and informed our codesign process. *Kilo* became centered in our codesign process for data science learning as a tool and language for the researchers, teachers, and students to share. We found connections between common traditional practices in science and data science and *kilo*. The *kilo* practices of 1) making observations, 2) forecasting and predicting, 3) fostering a relationship with nature, and 4) stewardship of the land and nature were connected to the pillars of community-centered data science that we identified across multiple layers of the humanistic framework. Our work enacted youth learning that was personal, cultural, and sociopolitical – each layer of the humanistic framework (Lee et al., 2021). Data science curriculum designed with community members and for a unique context, such as the work shared here, allows for the multiple layers of community connectedness to manifest. Students have varied and nuanced identities which come with them to school. Practicing *kilo* afforded youth a meaningful opportunity to bring their community funds of knowledge and identities into the classroom as valued and expert knowledge.

## Acknowledgements

The authors wish to acknowledge the generous support of the National Science Foundation of this work through NSF Grant #1942500. The opinions and findings herein are the authors and do not represent the NSF.

## References

Bang, M., & Medin, D. (2010). Cultural processes in science education: Supporting the navigation of multiple epistemologies. *Science Education, 94*(6), 1008–1026.

Bridges, D., & Bridges, D. (2017). "Nothing about us without us": The ethics of outsider research. *Philosophy in Educational Research: Epistemology, Ethics, Politics and Quality*, 341–361.

Calabrese Barton, A., & Tan, E. (2019). Designing for rightful presence in STEM: The role of making present practices. *Journal of the Learning Sciences, 28*(4–5), 616–658.

Calabrese Barton, A., & Tan, E. (2020). Beyond equity as inclusion: A framework of "rightful presence" for guiding justice-oriented studies in teaching and learning. *Educational Researcher, 49*(6), 433–440.

Carlone, H. B., & Johnson, A. (2007). Understanding the science experiences of successful women of color: Science identity as an analytic lens. *Journal of Research in Science Teaching, 44*(8), 1187–1218.

D'ignazio, C., & Klein, L. F. (2023). *Data feminism.* MIT press.

Gallimore, R., & Goldenberg, C. (2001). Analyzing cultural models and settings to connect minority achievement and school improvement research. *Educational Psychologist, 36*(1), 45–56.

Hawkman, A. M. (2020). Swimming in and through whiteness: Antiracism in social studies teacher education. *Theory and Research in Social Education, 48*(3), 403–430.

Kyza, E. A., & Agesilaou, A. (2022). Investigating the processes of teacher and researcher empowerment and learning in co-design settings. *Cognition and Instruction, 40*(1), 100–125.

Ladson-Billings, G. (1995). Toward a theory of culturally relevant pedagogy. *American Educational Research Journal, 32*(3), 465–491.

Ladson-Billings, G. (2021). Three decades of culturally relevant, responsive, and sustaining pedagogy: What lies ahead?. In *The Educational Forum* (Vol. 85, No. 4, pp. 351–354). Routledge.

Lee, V. R., Wilkerson, M. H., & Lanouette, K. (2021). A call for a humanistic stance toward K–12 data science education. *Educational Researcher, 50*(9), 664–672.

Manz, E. (2016). Examining evidence construction as the transformation of the material world into community knowledge. *Journal of Research in Science Teaching, 53*(7), 1113–1140.

Morishige, K., Andrade, P., Pascua, P., Steward, K., Cadiz, E., Kapono, L., & Chong, U. (2018). Nā kilo ʻĀina: Visions of biocultural restoration through indigenous relationships between people and place. *Sustainability, 10*(10), 3368.

Moscovici, S. (1998). The history and actuality of social representations. In U. Flick (Ed.), *The Psychology of the Social* (pp. 209–247). Cambridge University Press.

Owen, D. S. (2007). Toward a critical theory of whiteness. *Philosophy & Social Criticism, 33,* 203–222.

Titcomb, M. (1972). *Native use of fish in Hawaii* (Vol. 29). University of Hawaii Press.

Tofel-Grehl, C. (2023). "There is no room for me, for a Hawaiian, in science": Rightful presence in community science. *Journal of Research in Science Teaching, 60*(8), 1879–1911.

# 6

# HUMANISTIC PRE-SERVICE DATA SCIENCE TEACHER EDUCATION ACROSS THE DISCIPLINES

*Victor R. Lee*

## Introduction

Lee, Wilkerson, and Lanouette (2021) called for K-12 data science education, a rapidly growing area of interest for educators and policy makers at the time of this writing, to be viewed and approached with what they called a "humanistic stance". A humanistic stance centers human dimensions of data constructed from (1) direct encounters and experience of the phenomena represented by the data and (2) from operating in socially constructed, inherently cultural arrangements of meaning, routines, and power. A contrast to this would be viewing data and its use as absolute, determinative, transcendent, and removed from subjectivities of human design or interpretation. In that contrast, data would be discussed as self-evident, fully authoritative, and neutral. For example, were someone to assert that they were at an ideal body weight for their height because their body mass index (BMI) is 23.5 (i.e., within the recommended BMI of 18.5–24.9, according the Centers for Disease Control in the United States), a humanistic stance toward this would more intentionally acknowledge that BMI is a human-constructed measure that relies on a set of other human-constructed measurements (e.g., height as stable number – it changes during the day due to gravity; weight rather than other measurable body attributes such as muscle mass or body composition). Moreover, it is enmeshed in a milieu of human activities about what is socially constructed and maintained as ideals for what is "healthy" and how that is determined, which can and does vary. For instance, different nations have different ranges for classification as "normal" or "healthy" as do different individual people. It would also be recognized as having limits. It has different meanings and uses for different bodies, ranging

DOI: 10.4324/9781003364634-6

in variation from mobilities, sex, ancestry, bone density, and musculature, and life stage (i.e., childhood vs. adulthood). A humanistic stance is also mindful that how BMI is used – to identify someone as "underweight" or "obese" – has social meanings and consequences. These become bases for constructed ideals of beauty and judgments of individual attributes such as intellect, willpower, or even judgments about family and upbringing. It also becomes rationale for demanding certain changes in lifestyle (e.g., purchase or consume only certain foods or pay for membership at a health club; social permission to wear certain types of clothing to show or hide parts of the body).

To be clear, this is not to say that a humanistic stance requires the entirety of these complex social matters for the single BMI measure must be known by each person who uses it. It is not tractable nor practical to maintain nor advance that expectation. However, the humanistic stance Lee et al. recommends is that when working with data, we maintain awareness that this level of complexity exists, that there is importance and value to acknowledging these qualities, and there are times when those should be prioritized as part of data work. Also, a humanistic stance does not go to the extreme of absolute relativism. It does not assert that the social nature of any associated BMI data and acts of BMI datafication render all other qualities and features of the measure, such as those that are mathematical in nature, as devoid of meaning. Statistically, BMI can and does exhibit long-tailed distributional shapes when large samples and populations are plotted. It is indeed correlated with and can be predictive of some amount of variance in statistical models. And it also can be a very useful metric for many people. However, it – like all data and measures – is produced and operates in a complex social world and in one where we want data to be promoted and used more as a support rather than as a harm.

The operationalization of a humanistic stance as offered was in terms of layers of mediating relationships that are personal, cultural, and sociopolitical within a learning experience. There are ways to look at how data are used, learned, and taught that operate within and across these layers, and the individual naming of those as well as bundling of them together as under the "humanistic" label serves to offer a conceptual lens for researchers and educators to both critically and generatively think about the teaching and learning of data science. I note also that it is not a stance that had been newly "invented" by the authors as it is derived from and parallel to what others have done that appear in the literature and that appear in practice. Rather, it is a naming and bundling of ideas that are expressed by other scholars as well. Many other scholars have done work in much the same spirit, which includes but is not limited to work by Rubel et al. (2016) in their design and use of data experiences for students to critically examine neighborhoods; by the individuals described in Calabrese Barton et al. (2021) who were actively involved in the sensemaking and interrogation of "big" and "small" public data from the COVID-19 pandemic, in many examples summarized in the book *Data Feminism* by D'Ignazio and Klein (2020); in the critical examination of how racial harm can be embodied in and perpetuated through data science technologies (Benjamin, 2019), and many others.

With that brief overview established, the work of the chapter is a reflective examination of a pre-service teacher education course design based in this stance. It is not an authoritative compilation of design strategies, both because presenting this work in that manner would run counter to the premise of this chapter and because the field of pre-service teacher education related to the teaching of data science is young and still in need of more examples. However, the case upon which this chapter is based is one that had been very carefully and intentionally considered.

This chapter intentionally focuses on discussing the design of university-based teaching. It parallels similar chapters that discuss the design and shares vignettes from new university teaching experiences, such as one by Fields and Lee (2016) who reflected upon and shared design decisions and accounts of a university-based craft technology course in which studio practices, digital making, and artisanal craft were negotiated with one another in the creation of a university-situated learning experience. Indeed, critically analyzing the design of learning environments is arguably at the core of some areas of educational research.

### Context and Overview

This pre-service course was developed and offered at the same time public efforts to mobilize and advocate for data science education in K-12 were underway (2020 onwards), such as the creation and launch of *"Data Science 4 Everyone"* (datascience4everyone.org). That was also when the COVID-19 pandemic forced universities to conduct most of their teaching via remote learning, including the iteration of the course (offered Spring 2021) that is described here. However, the course was run similarly when in-person teaching resumed the following year (Spring 2022); the same topics and assignments were retained, and collaboration on documents in 2021 was instead done through physical materials in 2022 (i.e., a shared whiteboard rather than a shared online document).

The context for the course was the Stanford Teacher Education Program, a year-long teacher preparation program at Stanford University that guides accepted teacher candidates who already hold bachelor's degrees through credentialing, student teaching, and a combination of methods and theory coursework. The course was open to interested undergraduate students, many of whom are considering teaching, although it was explicitly communicated that the course experiences would be oriented toward matters of specific interest to teacher candidates. Two undergraduate students who were not yet pursuing a teaching credential opted to enroll with these caveats. For the remainder of the chapter, this course will be characterized as pre-service teacher education given its intent and majority enrollment.

This particular course on data science education was presented as an elective for secondary teaching candidates. It was unclear what level of curiosity or interest there was among the pre-service trainees, and policy conversations about data science in the curriculum were not yet in the public sphere. As such, it was expected to have a diverse draw from many different subject areas. This was both a constraint and motivation for the course, as described below.

Given growing interest from word of mouth about the existence of this course – which had been accelerated by independent promotion of the course by *Data Science 4 Everyone* – a decision was made to explore steps to make lessons learned about the course and share its design more widely. The disclosure of the activities of course, sample student work, and student written records underwent review by the university institutional review board and a separate independent committee that oversees and determines whether to allow for the use of Stanford student data for individual research projects. The consenting process was led and done by an unaffiliated and experienced education doctoral student with the instructor and teaching assistants removed from the space at the end of the course. No incentives were offered except for the potential knowledge that this disclosure may offer the field writ large, and no member of the instructional team was allowed to know who agreed to participate. That is, the instructional team was not permitted to know who gave permission for their class activity, as captured in digital artifacts and in the learning management system, to be shared anonymously until after the term had ended and grades had been submitted. Students had the option to request any specific items from the course be excluded.

Course enrollment was 15 students, including students who identified as male, female, and nonbinary, and was comprised of individuals of Asian/Asian-American, Latinx, and White students. Consent was provided by all students. As this was planned as a new course design and only later was recognized as having lessons and recommendations useful for other teacher educators, it was not designed with any form of pre-post metrics in mind. Thus, the empirical components are descriptive, with aggregated counts provided and specific cases and excerpts selected to give the reader a sense of breadth and specific instances.

### Students' Initial Expectations

Given the elevated recent interest in the topic of data science in educator spaces, the course began on the first day with an anonymous activity of all students contributing to a jam board with what they believed merited or motivated the idea of including data science in the K-12 classroom. Nineteen responses were posted with several receiving "+1" annotations to indicate another student's agreement or enthusiasm for a post, which was a practice of students in this teacher education programming with remote learning and online whiteboard usage. To give some synthesis, I offer the following code-and-count analysis: I took the 19 posts and reviewed them to infer common recurring themes, identified which one theme was the most prominent in the post, and then counted the occurrences. The "+1" annotations were ignored for this. Here is the breakdown, also provided in Table 6.1 for ease.

**TABLE 6.1** Themes in student posts for why it would be important to teach data science in K-12

| Theme | Description of Theme | Count (N = 19) | Examples |
|---|---|---|---|
| Abundance | The tremendous amount of data in the world and that are being generated currently motivate the need to learn data science | 8 (42.1%) | • There is more data being created today than ever before and it's growing exponentially!<br>• Because students are overwhelmed with data all day everyday |
| Intellectual proficiency | The ability to understand and interpret data is inherently valuable for comprehension, inference, and knowledge building | 5 (26.3%) | • We must teach kids to be able to recognize trends and patterns in the real world<br>• this will equip students with science literacy - so much of science is hard to analyze if you don't know how to read and analyze/interpret data |
| Economic | The ability to know and practice data science is important for employment and earnings potential | 3 (15.8%) | • We need to prepare students for the future hi-tech economy |
| Equity | The use and practice of data science have potentially harmful social outcomes and the exclusion of historically marginalized social groups implicated | 2 (10.5%) | • to avoid bias in data solutions we need a diverse team of data scientists… this starts with all students receiving access to data science. Education [sic] |
| Empowerment[1] | The content and tools of data science can enrich and augmented valued individual strengths | 1 (5.2%) | • We must teach data science because it is a field that channels a lot of self-reflection, views and observations which is an area that kids thrive in! They are so creative! |

The most prominent theme was that data are now pervasive. The motivation was based on abundance of data, with some specification that it was in many sectors of our lives. This was the dominant theme in eight posts (42.1%). Posts speaking to this include "There is more data being created today than ever before and it's growing exponentially!" and "Because students are overwhelmed with data all day everyday". The second most frequent was that data science education would be related to a valued intellectual proficiency. Here, the ability to work with and interpret data was broadly useful for understanding or knowing things in the world when data were involved. This appeared in five posts (26.3%) and is represented by statements like "We must teach kids to be able to recognize trends and patterns in the real world" and "this will equip students with science literacy – so much of science is hard to analyze if you don't know how to read and analyze/interpret data". The other identified themes related to motives for teaching data science included: (1) economic – with employment or income being named – which had three posts (15.8%) (e.g., "We need to prepare students for the future hi-tech economy"); (2) equity – related to addressing potentially harmful social outcomes of data science – which appeared in two posts (10.5%) (e.g., "to avoid bias in data solutions we need a diverse team of data scientists...this starts with all students receiving access to data science. Education [sic]"); and (3) student empowerment – with only one post (5.3%) that did not tie to empowerment related to economic or equity concerns (i.e., "We must teach data science because it is a field that channels a lot of self-reflection, views, and observations which is an area that kids thrive in! They are so creative!").

In all, there were a range of concerns represented, with the sheer amount and pervasiveness of data being highly represented. This implies that in part, by virtue of many people – as represented by media and other outlets – were talking about it, it was therefore an important topic to teach. Of note also is the concern expressed by the student who commented about the role of data analysis and interpretation in science (the *Next Generation Science Standards* have that as a specific science and engineering practice to support in K-12). I understand this as a recognition that data had already been a topic of interest in other subject areas and thus a redoubled focus on teaching data and data science was merited. Implicit is that given data science was influencing multiple forms of practice, science (and science education) could benefit from the focus as scientific practice and knowledge generation were/are changing in light of the ability to utilize massive amounts of data and data science analysis techniques. Another post was considering how literacy was taking on new meanings and data would be implicated: it is/will be a component of "literacy". Equity issues with respect to who is given more resources to study data science or who may be harmed in the use of data science were mentioned but not as foregrounded. The individuals who raised this were noting that the ability to use and participate in the data scientific work was a means of wielding power and influence, whether it was economic opportunity or the ability to influence others. Even for those who see themselves as informed and well-intentioned, this also represents a call (see Table 6.1) for diverse perspectives and lived experiences to have access and influence at major decision-making points and in for a plurality of individuals to receive and experience purported benefits of data science. However, this was not frequently named (nor were these post that received any "+1" marks from other students). This is not to say such concerns were unknown to the students, but they were less readily volunteered nor flagged as key motivators for teaching data science.

### Orientation toward What Is Data Science Content

Currently, data science is often mentioned and elevated as important, but what data science comprises is a bit unclear. For some, it looks very similar to what had historically been called statistics (e.g., Gould, 2021; Rubin, 2020). For others, it is very intentionally an intersection of statistics and computation (in the form of programming). Others present it as a way of thinking about problems that involve data. Also prominent in broader discourses is that data science is comprised of a very specific set of computational techniques for working with large datasets. To a surprising amount, a blog post (with the "Conway Venn Diagram") has shaped much of the discourse of what is and is not data science although ambiguity remains on boundaries and what is included and appropriate in K-12 (Lee and Delaney, 2022). A broadly accepted definition is likely to remain elusive for some time, although I do see value with all of these. This led to a tension of how to represent data science for pre-service teachers.

The decision for this course was to acknowledge that data science is often associated with specific techniques and activities involving machine learning, programming, and statistics, but it would be unrealistic in the course of a single quarter to expect mastery for pre-service teacher candidates with other pre-service teaching obligations who may have had very limited, if any, prior formal training in computer science, programming, or statistics. At the same time, and in the spirit of Bruner (1960) who asserted that any subject could "be taught effectively in some *intellectually honest* form to any child at any stage of development" (p. 30), my desire in offering this course was to provide more onramps for pre-service teachers. Perhaps they would not leave the course knowing the names and trade-offs of different clustering algorithms, but they would be able to see accessible forms of data practices that are access points for more technical data science work. I did not seek out to teach anyone to jump into industry as a data scientist, but I wanted them to feel knowledgeable of foundational ideas, what fields tend to offer related content, and robust awareness of how increasingly present data science was in many aspects of their modern lives.

With that said, I did intentionally decide early in the term to commit an entire class session to sharing how data science was instantiated professionally. This included a brief historical overview of artificial intelligence, its stagnation (i.e., "AI Winter"), and the renaissance in neural network and "deep learning" paradigms given new data availability. The distinction between supervised and unsupervised machine learning was introduced, actual job listings in industry for "data scientist" positions were examined to identify and discuss specific terminology, and a python notebook with a simple three-layer neural network was shown and run. Additional resources were provided such as the *Tensor Flow Neural Network Playground* (https://playground.tensorflow.org/) to show interactive visualizations of what neural networks look like and how manipulating (training) them could yield different detection capabilities on a two-dimensional graphical representations. A couple of students who did have computer science backgrounds were enthusiastic participants. However, admittedly it was challenging content, with a course evaluation comment being "I am still not sure about the more technical, computer science specific parts of gathering and processing data, though. I feel like I needed an introductory course for those components". Aside from a single week exploring these components of data science, the only other appearance of computer programming was in examination of existing secondary school curriculum materials for data science courses, but that was done in passing. With those technical aspects represented, the remainder of the course took on other emphases, described below.

### A Humanistic Framing of Major Design Decisions

In several respects, the design of the course tracked with what is common in the design of many university-based courses. This included identifying what primary course goal to emphasize (which ultimately was for students to produce their own lessons and instructional activities they could use for student data exploration), what course tools would be used, and how to structure activity types to support course learning goals. With that said, there were three major decisions that were intentionally made in the organization, scoping, and sequencing that are described here with a humanistic stance in mind. Each of the three humanistic layers (personal, cultural, and sociopolitical) is discussed with a set of design decisions. The order is presented as personal, sociopolitical, and cultural to align with the sequencing of the courses.

### Personal: Dear Data Meets the Quantified Self

A very deliberate decision was made to have students begin the course with a multiweek personal data project of their own choosing. This was to create an intentionally and uniquely personal experience for students with data that could be referenced throughout the term. The project was inspired by the *Dear Data* book and the Quantified Self hobbyist community. *Dear Data* is a book (Lupi and Posavec, 2016) that graphically chronicles a year's worth of weekly exchanges between two graphic designer friends located on either side of the Atlantic Ocean. Each week, they would produce a hand-drawn visualization of some data capture of the week. Examples might include how many doors each of the authors walked through or how many complaints they made throughout the week. These were rendered on postcards with the data visualization on one side and a legend on the other. Aside from being aesthetically intriguing, as a variety of ways of using shape, color, position, and orientation were used by each and throughout the year, the simultaneous documentation by both authors each week of the same topic but different decisions allowed both detailed individual inspection and cross-author comparisons for patterns, similarities, and differences. It has become popular enough of a published collection that instructions are provided with the book and accompanying online media explaining how to emulate it. *Dear Data* has served as inspiration for other existing data science education research projects, such as one that sought to emphasize and explore intersections of art and data for cross-subject collaborations in schools (Matuk et al., 2021).

While *Dear Data* helped popularize the practice of capturing and displaying data from routine activities and experiences, there are noted examples that precede this that were practiced in smaller communities. One is the quantified self-hobbyist community, which has been studied separately (Choe et al., 2014; Lee, 2014). The term "quantified self" was popularized by Wolf and Kelley in 2007 (Wolf, 2010 summarizes some of the history of the term) in reference to the rapidly growing ability for anyone to easily gain data from and about themselves due to increased availability of sensor technologies. It is used to refer to the orientation of quantification of the personal and mundane, but it also refers to a community that formed across the globe that convened local meetup groups, hosted an annual international conference, and a community web presence with discussion boards and resources. Some research laboratories even formed around this interest and from the community inspiration. Participation in the community appeared to have heavy representation from one demographic (adult white males) (Lee, 2014), but a number of online materials and statements from key figures suggest that the quantified self was intended to be an open idea for anyone. The recommendation for a quantified self "Show and tell" presentation was to share answers to three specific questions to a public audience, whether it was a small community meetup, at the conference, or in videos (often accompanied by slides). These questions were:

- What Did You Do?
- How Did You Do It?
- What Did You Learn?

For this pre-service course, *Dear Data* and Quantified Self were integrated as an individually designed but required activity. Students were tasked with tracking 7–10 days' worth of some selected activity, design a visualization, and respond to the three quantified self "show and tell" questions. Unlike the *Dear Data* postcard project, they were not restricted in medium nor size of visualization (they could draw, use computers, or introduce any media of their choosing). Unlike a quantified self "show and tell", there was no formal talk/presentation expected, although students were invited to share. They had ultimate choice of what to track and disclose.

The pedagogical intentions with the activity were to expose students immediately to the challenges and decisions that are involved in constructing data, exploring how data can be represented in numerous ways, situate data in personally relevant contexts, and encourage articulation of how data were collected and recognize what inferences data visualizations can and cannot support under different conditions. In the class of 15 students, each designing and pursuing their own projects, we ended with three projects about pets (such as how often the dog lost their tennis ball and the student had to retrieve it for them), six about personal behaviors (such as when and where the student bit their lip), two about food or drink consumption (such as what liquids the student consumed), two related to mobile devices (such as text messages exchanged), one about scheduled activities, and one about personal mood or feelings. Two are provided below as illustrations. In Figure 6.1, Trevor (pseudonym) plotted his journeys into Wikipedia. As he described it:

> I LOVE Wikipedia! It is a constant source of discovery and exploration. I'm on Wikipedia at least once a day, usually spurred by something I see or hear about and want to learn more about, and then usually click on a couple pages when I'm like "Oooh, I wonder who that person is, etc."

Through this activity, he coded his data based on categories and noted one major learning was how many potential ways one could categorize data and that one can have biases in what data categories they pick. Trevor also noted he could not find a pattern with respect to the categories he selected in the "chains" of searches, which was sanctioned as an important and legitimate finding. This activity allows students to discover that in any investigation, data can fail to yield clear patterns. In textbook mathematics and science classroom activities, this is not a message that is frequently communicated to students.

Zayn's data project (Figure 6.2) was to examine how much trash they produced and discarded over seven days. Zayn lived in campus residential housing and spent much of their time at campus sites. A skilled programmer who was also completing a math degree, Zayn opted to base their visualization on a

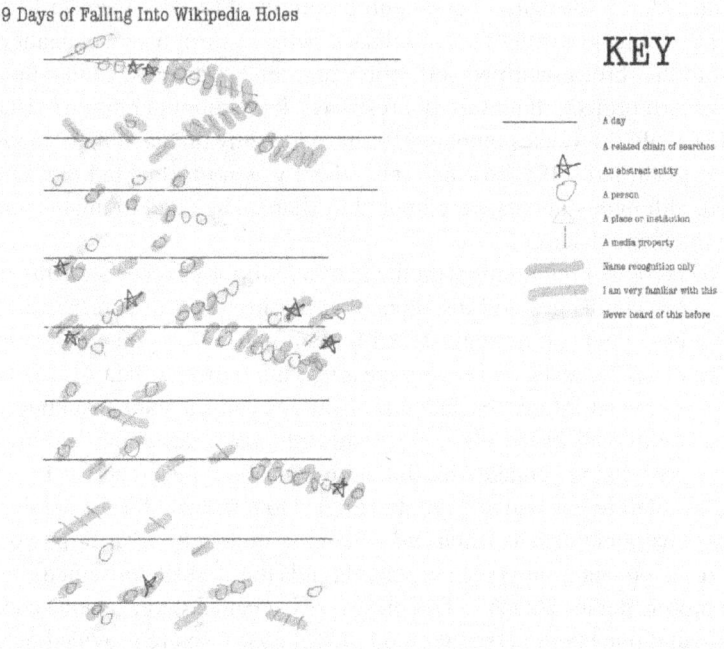

**FIGURE 6.1** Trevor's visualization of his Wikipedia Reading behavior tracked over nine days.

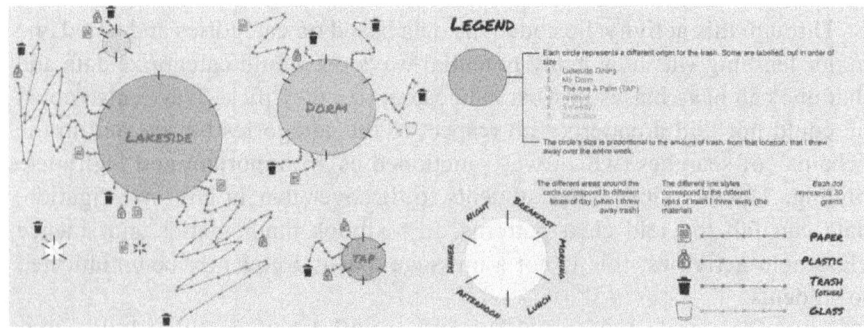

**FIGURE 6.2** Zayn's data visualization of their trash production and disposal over seven days.

mathematical model and render this through a program they wrote. This effort at precision had some challenges, shared below.

> Precise data visualization was also a bit of a pain. I used math to make sure that each of the areas of the circle is accurate, given the amount of trash I threw away at each location. This became a problem when it came to generating the tendrils for each of the circles. Each tendril is modelled [sic] by the function so I was trying to generate subplots of that function where the length of the subplot corresponded to the amount of that kind of trash I threw away. This turned out to be very complex because I had to calculate the arc length of that function... in the end, I just used a linear approximation for the length of the function, so they're not quite precise and the dots on the tendrils aren't spaced out evenly along the tendril.

Discoveries, in addition to the limits of various computational and mathematical choices for manipulating the data, also included the amount of waste that they produced. Surprised that so much of the waste that they had was from their residential dining hall, Zayn explored publicly available data to estimate how much waste was being produced by all students using the dining halls, to the tune of 10,000 kg per week. This was a cause for concern about waste and environmental responsibility for a major institution of higher education.

Many other projects were creative and compelling and rife with discoveries about collecting, structuring, visualizing, and interpreting data. Students were able to recognize shortcomings in their processes and important factors not captured in their data. Some were equipped to make personal discoveries of interest, such as how drinking tea was linked to socializing (on a project cleverly titled by the student "What's the tea, sis?" to represent tea consumption and the slang for the sharing of gossip or updates about other people) or in another project, how time of day was associated with higher and lower amounts of fidgeting.

Aside from the individual discoveries students made and appreciation of what human actions are associated with the production and processing of data, this activity served to introduce points that were continuously revisited and referenced in the class. These included that some information is always missing in data and that how data were structured was a choice.

### Sociopolitical: Early and Recurrent Examination of Harms, Risks, and Exclusions

In Table 6.1, we saw that "equity" was not the primary reason why students thought data science should be taught in K-12. However, it was and is a topic of interest to the instructional team given known potential for (and several already documented cases of) societal harm with data science. Therefore, addressing this was a priority and motivation for the course proposal and design. With that said, matters of equity, harm, and exclusion in data science are very complicated and unresolved topics. The work needed to deeply address these is substantial; I do not claim what was done in this course is superior to other options, nor necessarily even sufficient to address such matters given their complexity. With that said, my goal in this section is to share specific strategies, decisions, and actions in this course. My broader hope is that others can take these examples from here and further develop them and make them better – or produce superior alternatives. To keep this a consistent and persistent focus, the course was designed to repeatedly examine harms and risks and to keep asking questions about who does and does not hold positions of power with respect to what happens in the production and use of data science.

One action taken in course design to uphold this commitment was taken respect to sequencing. It is not uncommon for matters of diversity or equity to appear "appended" to courses that are not explicitly described as being about the explicit topics of diversity and equity. In the third week, the topics of bias, equity, and ethical matters such as privacy, were made the main foci. By having this in the early parts of the course, the intention was to maintain this as a lens for the remainder of the course and revisit these matters as we discussed examples of instruction and educational resources in the remaining seven weeks. In the interest of promoting continuity, the course progressed with regular reference to previous weeks' content, making this early positioning a small, but intentional action. That is, the same topics from earlier in the course were deliberately revisited as new topics were introduced.

To make sociopolitical concerns relevant, the instructional team decided to provide special emphasis on data privacy. This is a major and current sociopolitical concern given that surveillance capitalism operates on the continuous capture of our data. For example, one slide from the class shown in Figure 6.3 shows the ranges of everyday activities that do not obviously present themselves as pertaining to data but actually are dependent on continual production and sharing of our data to operate and yield profit. This includes when labeling images in CAPTCHAs, when marking an online review as helpful in decision-making, and when scanning supermarket member cards to get discounts on products (and to maintain purchasing history for individuals for marketing and advertising).

Another way in which we sought to surface sociopolitical issues was to create situations where students ask questions of inclusion and representation. For instance, I sought to surprise some students about whose work with data gets recognized in the public canon and whose does not. Class content included in-class review of some of W.E.B. Du Bois's work with data infographics (Battle-Baptiste and Rusert, 2018) to make explicit that important and innovative data

# What all is producing/capturing the data?

• Apps, devices, websites

**FIGURE 6.3**    Examples discussed in class of everyday instances where data are being produced and captured that yield profit for outside entities.

visualization work was done by Black thought leaders well before data science became a hub for data visualization conversations. Prior to its release on streaming media services, the *Coded Bias* documentary (Kantayya, 2020) which documents cases of algorithmic bias and failures of data-trained technologies to respond to Black users (e.g., Joy Buolomwini's work with face recognition technology showing how commercial technologies failed to recognize her face with a darker complexion while readily recognizing it when she donned a white plastic mask is one noted example) was specifically pre-arranged to be available for all students to stream through university library media resources. Cases of harm and exclusion described in books such as *Weapons of Math Destruction* by Cathy O'Neill and *Algorithms of Oppression* by Safiya Noble were presented and discussed. Examples of data as a tool for exposing and critiquing unjust systems and expanding discourses from *Data Feminism* were similarly presented and discussed. Several examples were new to students and led them to comment on how major tech companies (many of which were headquartered nearby) heavily employed males of certain trainings and backgrounds. The "tech bro" term was raised, and the associated disposition of that caricature was named by students as one factor leading to these instances of harm and exclusion. Patriarchy and entrenched sexism also surfaced in discussions of examples from *Data Feminism*.

Opportunities to examine actual loci for concerns related to harm and bias were also woven into the course. Students were intentionally introduced to pedagogically oriented data tools and platforms that they could use in their own teaching, such as the free CODAP data analysis platform and license-based services such as *Tuva* or *DataClassroom*. Beyond illustrating how these platforms worked with their existing sample datasets, a specific dataset to speak to current sociopolitical concerns was also provided in .csv format for students to analyze: public policing data from the *Open Policing Project* (https://openpolicing.stanford.edu). At the time this course was designed and implemented, racial justice protests and subsequent statements and actions were being taken in response to rising awareness of police brutality and harms done to Black individuals and communities. Interest was high nationally, and the class engaged in discussions with random subsets of policing data from specific cities to explore proportions of arrests and traffic stops and to raise questions about what is and is not shown in these data. For instance, groups identified cases of Black individuals being disproportionally (based on relative population in the cities providing data) being stopped and cited in police records. They also observed that these datasets would not necessarily include when someone was "let off with a warning" and that the numbers and words in a spreadsheets and .csv files are very different types of records than those coming from dashboard or body camera footage that some metropolitan areas and municipalities have required for law enforcement officers.

### Cultural: Emphasis on Multidisciplinary

Education researchers have been making persistent calls for decades to understand human activity, learning, and development as inherently and fundamentally cultural (Nasir et al., 2020; Rogoff, 2003). This means that the tools, language, and routines that are taught are connected to purposes and practices that coordinate the activities of multiple individuals. Norms, goals, and rules operate and persist sometimes invisibly with some continual social negotiation for adaptation and change over time. Culture shapes what people do, but then what people do because of cultural forces also serves to reshape culture. Given that orientation, learning in disciplines can be understood as culturally varied in that each discipline has specific tools, discourses, norms, and goals.

Data science, if understood as being more than a standalone domain, interacts with disciplinary cultures in complex ways. Existing practices of the discipline shape what value and role data science should play, but at the same time the ready access of data create new opportunities for newcomers to participate in the practices of a discipline. Elsewhere, some colleagues and I (Jiang, Lee, and Rosenberg, 2022) have called for actively exploring K-12 data science education as it exists across disciplines so that data science proficiency can be better situated, integrated pragmatically into the curriculum, and so that more access to data science and the disciplines can be provided given new pathways facilitated by new data practices. That orientation had been at the core of this preservice course design. As such, two weeks were explicitly dedicated to examinations of how data abundance take shape in social studies, the humanities, art, and the natural sciences.

From the activities, public examples, and assigned readings, the following observations were made. For social studies, maps, language, and human relationships were stressed as major focus areas and ways to integrate data with existing lines of inquiry. For instance, the class looked at census data in geographic data visualization systems (such as *Social Explorer*) and read examples such as Radinsky et al. (2014) to see how map representations play powerful roles in social studies data work. Language and large text corpora were flagged as examples in the class for how digital humanities can explore changes and variations in how humans communicate under different circumstances, such as over time or in different geographies. Human relationships, beyond those latent in the examples above, were explored in class through examples such as the status of male and female characters in film based on amount of speaking time or dating tendencies as shared in dating service data.

For arts, novel data visualizations and efforts to link data with artistic expression such as those described by Matuk et al. (2021) were examined. Additionally, visualizations from the online longform data journalism publication *The Pudding* (at pudding.cool) were examined in small groups. Examples included comparisons of word usage by rap artists and in relationships and recurring themes in the lyrics of the musical *Hamilton*. Data journalism and public faces of data were also explored for the sciences, with science infographics and recommended pedagogical practices from Lamb et al. (2014) shared and discussed during class exercises reviewing data infographics together. Citizen science platforms, such as *GLOBE Observer* and *FieldScope*, as well as associated lesson plans were reviewed, with the recognition that citizen science is one contact point for learning experiences with large datasets (Harris et al., 2020). Additionally, resources such as *Data Nuggets* (described in Shultheis and Khelvik, 2015), which provide lesson materials using data obtained from professional science research studies, were examined as a curricular resource as well. In both arts and sciences, the ways in which the respective disciplinary cultures were represented was a point of discussion. For instance, arts can serve to examine and represent the human experience, and the class discussed how data tools supported that and how artistic practices could change ways we represent and relate to data. For science, the ways in which data are used as evidence in forms of argument and persuasion were discussed and examined.

### Student Products from the Course

A culmination for the class was the custom creation of data-intensive lessons for subject matter that the teacher candidates expected to teach in the future. Decisions and context for this course have been shared, but I cannot make strong claims about learning. This was not a systematic research inquiry with pre-developed measures and much of the data was anonymous to me (such as the Jamboard activity summarized in Table 6.1), so I could not trace trajectories or changes for specific individuals. However, I can examine lightly if there was some evidence that supports the notion that the course influenced teacher candidates' thinking about instruction. In service of this, I summarize an example from an English teaching candidate and from a Science teaching candidate.

The English teaching candidate's lesson sought to engage high school English students in an examination of book and author popularity data to support claims about what literatures are valued culturally and academically. A key data resource this teacher candidate planned to use was a set of public scatter plots of large datasets prepared by the Stanford Literary Lab (https://litlab.stanford.edu/LiteraryLabPamphlet17.pdf). These plots were made using data from the International MLA bibliography, which compiles academic literary criticism and thus shows how often authors are the topic of published literary critic engagement, and ratings on the *GoodReads* book rating platform, which shows one measure of popular uptake. The teacher candidate wrote the lesson plan such that students would work in small groups to review some of these data and analysis. Then the teacher candidate would open discussion and student interrogation into what counts as prestige in literature and who gets to make that determination. This culminated in students participating in a debate as to what is represented in "the canon" of quality literature. To close the lesson, students completed an exit ticket where they stated what they thought would be their own ideal criteria for texts in an English curriculum. This activity involved large datasets but already rendered into plots and had points of personal, cultural, and sociopolitical contact for students.

The science teacher candidate opted to focus on data infographics for their lesson. Interestingly, they opened with class examinations of some of WEB Du Bois's Infographics and then incorporated Lamb et al.'s (2014) guidance of think-aloud interpretations of Du Bois' infographics followed by a group think-aloud a greenhouse gas emissions infographic the teacher candidate had found online. The class would then be explicitly oriented toward thinking about how climate change is represented and use the *Data Classroom* platform and sea level datasets to examine for changes over time. Using *Data Classroom*'s tools, the students would then export plots to create their own infographic about climate change. The next day, students would then look at infographics related to climate change denial and how data interpretations would be recrafted intentionally to advance climate denialism. Then students would finish their own designed infographics, present them, and they would then be mounted around the classroom. Here, there are interesting and some unexpected points of contact with the preservice course design decisions described above. For example, including Du Bois's infographics for an environmental science lesson was unanticipated but sensible in the context of orienting students toward interpreting infographics and making visible to students that infographics have a long and political history. The approach of having students look at infographics from the lens of climate change denial was an intriguing one to further students' understanding of complex current sociopolitical issues. Science standards alignment was strong with appropriately related key performance expectations and associated science and engineering practices, cross-cutting concepts, and disciplinary core ideas from the Next Generation Science Standards stated in the lesson plan.

While these are just two examples, they do suggest that personal, cultural, and sociopolitical dimensions – which had not been explicitly called for in those terms during the course – had indeed been taken up by students.

## Reflections and Closing

With data science education in K-12 continuing its forward march, there will undoubtedly be more preservice teacher preparation efforts to address this content. This could take the form of entire dedicated courses like the one summarized above, but given the already noted heavy requirements for credentialing and certification of teachers, it may be that data science is embedded in other methods preparation. Mathematics teaching methods might be one home for it, and it is important to acknowledge that key ideas of data science and statistics teaching have long been emphasized by thoughtful teacher educators before data science became "trendy". However, for reasons enumerated in Jiang et al. (2022), I hesitate to suggest mathematics should be the sole home for data science education. Another home could be in computer science teaching methods, although the paucity of students pursuing pre-service certification in computer science teaching (see National Academies of Science, Engineering, and Medicine, 2021), especially relative to other longstanding subject areas, does raise questions about how broadly such preparation will reach. Integration across pre-service teaching methods courses may be another avenue, although there will be challenges given the numerous extant requirements for preparation of teachers in specific subjects and disciplines. Where and how such emphasis would ultimately fit is a harder question to answer that is beyond the scope of this chapter. It involves reflection on what should be in the curriculum and for what purposes. It also involves complicated matters of politics and stakeholder interests, for which others undoubtedly have more expertise.

However, for the conceptualization of course experiences to represent key values and presuppositions about the nature of an emerging discipline – such as those represented in Lee, Wilkerson, and Lanouette's (2021) humanistic stance – I have provided one example. It was one early attempt where others could certainly do more and make different decisions. But it should illustrate how a specific lens on data science education could be operationalized in the structure and enactment of a pre-service teacher education course. There is also, within the described course, room for improvement. Some students wanted to primarily learn how to plan data science relevant lessons whereas others wanted to know what data science was and others still wanted to learn how to teach very specific curricula and lessons. Addressing all of these needs adequately is a challenging constraint satisfaction problem, and decisions about which of them are emphasized most in such a course would need to take into consideration time constraints (e.g., a 10-week quarter is markedly shorter than a 14-week semester, for instance – and a course offered in the final term of a pre-service preparation program when edTPA, student teaching, and applications for teaching positions were primary concerns for me and may require different design decisions than one done at the beginning of a teacher preparation program). However, even if a reader sees this and opts to pass on specific design decisions, we have reason to believe our actions as teacher educators can be impactful. The inclusion of WEB Du Bois's data infographics seeing that this one example appeared in a science teacher candidate's designed lesson suggests it had some value. If indeed that one teacher candidate, now a practicing teacher, did use this in their own teaching, then they in turn engaged a number of students who will now have this as part of their data learning experiences. With time, this could yield large downstream effects, in the way that some theorists would suggest the flapping of butterfly wings could impact the formation and trajectory of a tornado. The sharing of this single course design within this chapter is also a very small action, but perhaps as more readers encounter the ideas shared here and explore and improve upon them through their own teaching, we might see a humanistic imprint on data science education in the years to come.

### Positionality Statement

The author identifies as an Asian-American cisgender male child of immigrants who has professionally favored explorations of STEM teaching and learning from the perspectives encouraged in the learning sciences – this includes valuing theories, methods, and forms of argument used in cognitive science, design sciences, and social research but with a sense of some accountability for ideas to speak in some measure to the realities of complex learning systems. He was the primary designer and instructor of the course and had embarked on this teaching effort without expecting it to be discussed publicly and reported. For this course, two enthusiastic doctoral students, Daniel Pimentel and Tanya LaMar, participated as teaching assistants and were involved in helping to refine the thinking of the course design and support students. He is a professor of education at a well-resourced private university that grants a great deal of faculty autonomy.

## Note

1 While this was only coded as appearing once, it did have an additional +1 label placed by another student suggesting it was similar to what another student was going to write or cued strong agreement from at least one other student, therefore it is included in this table.

## References

Battle-Baptiste, W., & Rusert, B. (2018). *WEB Du Bois's data portraits: Visualizing black America*. Chronicle Books.

Benjamin, R. (2019). *Race after technology: Abolitionist tools for the New JIM Code*. Polity Books.

Biswas, G., Leelawong, K., Schwartz, D., & Vye, N., & The Teachable Agents Group-V. (2005). Learning by teaching: A new agent paradigm for educational software. *Applied Artificial Intelligence, 19*(3–4), 363–392. https://doi.org/10.1080/08839510590910200

Brown, M., & Edelson, D. (2003). Design Brief – *Teaching as design: Can we better understand the ways in which teachers use materials so we can better design materials to support their change in practice?* Northwestern University.

Bruner, J. S. (1960). *The process of education*. Harvard University Press.

Calabrese Barton, A., Greenberg, D., Turner, C., Riter, D., Perez, M., Tasker, T., & Davis, E. A. (2021). Youth critical data practices in the COVID-19 multipandemic. *AERA Open, 7*, 23328584211041631. https://doi.org/10.1177/23328584211041631

Choe, E. K., Lee, N. B., Lee, B., Pratt, W., & Kientz, J. A. (2014). Understanding quantified-selfers' practices in collecting and exploring personal data. In *Proceedings of the SIGCHI conference on human factors in computing systems*.

D'Ignazio, C., & Klein, L. F. (2020). *Data feminism*. MIT Press.

Fields, D. A., & Lee, V. R. (2016). Craft technologies 101: Bringing making to higher education. In K. Peppler, E. Halverson, & Y. Kafai (Eds.), *Makeology* (Vol. 1, pp. 121–137). Routledge.

Gould, R. (2021). Toward data-scientific thinking. *Teaching Statistics, 43*(S1), S11–S22. https://doi.org/10.1111/test.12267

Harris, E. M., Dixon, C. G. H., Bird, E. B., & Ballard, H. L. (2020). For science and self: Youth interactions with data in community and citizen science. *Journal of the Learning Sciences, 29*(2), 224–263. https://doi.org/10.1080/10508406.2019.1693379

Jiang, S., Lee, V. R., & Rosenberg, J. M. (2022). Data science education across the disciplines: Underexamined opportunities for K-12 innovation. *British Journal of Educational Technology, 53*(2), 1073–1079. https://doi.org/10.1111/bjet.13258

Kantayya, S. (2020). *Coded Bias* [film]. 7th Empire Media, Ford Foundation - Just Films.

Lamb, G. R., Polman, J. L., Newman, A., & Smith, C. G. (2014). Science news infographics: Teaching students to gather, interpret, and present information graphically. *The Science Teacher, 81*(3), 25.

Lee, V. R. (2014). What's happening in the quantified self movement? In J. L. Polman, E. A. Kyza, D. K. O'Neill, I. Tabak, W. R. Penuel, A. S. Jurow, K. O'Connor, T. Lee, L. D'Amico (Ed.), *Learning and becoming in practice: The international conference of the learning sciences (ICLS) 2014* (Vol. 2, pp. 1032–1036). ISLS.

Lee, V. R., & Delaney, V. (2022). Identifying the content, lesson structure, and data use within pre-collegiate data science curricula. *Journal of Science Education and Technology*, *31*, 81–98. https://doi.org/10.1007/s10956-021-09932-1

Lee, V. R., Wilkerson, M. H., & Lanouette, K. (2021). A call for a humanistic stance toward K-12 data science education. *Educational Researcher*, *50*(9), 664–672. https://doi.org/10.3102/0013189X211048810

Lupi, G., & Posavec, S. (2016). *Dear data*. Princeton Architectural Press.

Matuk, C., DesPortes, K., Amato, A., Silander, M., Vacca, R., Vasudevan, V., & Woods, P. J. (2021). Challenges and opportunities in teaching and learning data literacy through art. In In de Vries, Y. Hod, & J. Ahn (Eds.), *15th international conference of the learning sciences (ICLS)* (pp. 681–684). ISLS.

Nasir, N. S., Lee, C. D., Pea, R. D., & McKinney de Royston, M. (2020). *Handbook of the cultural foundations of learning*. Routledge London.

National Academies of Sciences Engineering and Medicine. (2021). *Cultivating interest and competencies in computing: Authentic experiences and design factors*. National Academies Press.

Radinsky, J., Hospelhorn, E., Melendez, J. W., Riel, J., & Washington, S. (2014). Teaching American migrations with GIS census webmaps: A modified "backwards design" approach in middle-school and college classrooms. *The Journal of Social Studies Research*, *38*(3), 143–158.

Rogoff, B. (2003). *The cultural nature of human development*. Oxford University Press.

Rubel, L. H., Lim, V. Y., Hall-Wieckert, M., & Sullivan, M. (2016). Teaching mathematics for spatial justice: An investigation of the lottery. *Cognition and Instruction*, *34*(1), 1–26. https://doi.org/10.1080/07370008.2015.1118691

Rubin, A. (2020). Learning to reason with data: How did we get here and what do we know? *Journal of the Learning Sciences*, *20*(1), 154–164. https://doi.org/10.1080/10508406.2019.1705665

Schultheis, E. H., & Kjelvik, M. K. (2015). Data nuggets: Bringing real data into the classroom to unearth Students' quantitative & Inquiry skills. *The American Biology Teacher*, *77*(1), 19–29. https://doi.org/10.1525/abt.2015.77.1.4

Wolf, G. (2010). The data-driven life. *New York Times Magazine*. http://www.nytimes.com/2010/05/02/magazine/02self-measurement-t.html?_r=1&ref=magazine

# 7

# EVERYDAY EQUITABLE DATA LITERACY IS BEST IN SOCIAL STUDIES

## STEM Can't Do What We Can Do

*Tamara L. Shreiner and Mark Guzdial*

In the wake of COVID-19 and amid concerns over misinformation and disinformation spreading through the internet and social media outlets, there were more calls for teaching data science and data literacy to all K-12 students (Drozda, Johnstone, and Van Horne, 2022; LaMar and Boaler, 2021; Levitt, 2022; Weiland and Engledowl, 2022). Such calls have been underscored by research indicating that students across the country fail to evaluate online information for validity and bias and that many students struggle to read even the simplest visual representations of data (Breakstone et al., 2021; Kahne and Bowyer, 2017; Konold et al., 2015; LaMar and Boaler, 2021). Many proponents of data science and literacy in schools focus primarily on infusing it more deliberately into mathematics (e.g., LaMar and Boaler, 2021; Levitt, 2022) or computer science (CS) education (e.g., Yongpradit, Hendrickson, and Phillips, 2016) to provide students with the hard skills they will need to make sense of and solve problems with data and data visualizations.

While we agree that teaching about data and data visualizations in these disciplines is important, we also believe that *social studies classes* should be a primary venue for teaching data literacy. As this chapter will argue, teaching data literacy in social studies provides opportunities to teach or reinforce some fundamental data literacy skills to all students while also teaching social studies content. More importantly, most people's everyday need for data literacy is in a social studies context where the purpose of instruction is to "help young people develop the ability to make informed and reasoned decisions for the public good as citizens of a culturally diverse, democratic society" (National Council for the Social Studies, 2021). While some people

DOI: 10.4324/9781003364634-7

need to interpret a scatterplot in *Nature*, most people try to gather information relevant to civic and social lives from infographics in *USA Today* or *The New York Times*. Social studies teachers are presumably equipped through their training and background knowledge as social studies teachers to teach equity-driven, justice-oriented data literacy that is too often missing from curricula in other subjects. As a recent landscape report on CS teaching argues, only 40% of teachers cover impacts of computing and only 42% cover ethics (Koshy et al., 2022). Teaching data literacy in social studies can address both topics.

However, there are challenges to making this work. First, we acknowledge that social studies teachers are not often properly prepared to teach about data and data visualizations and that not all social studies teachers may be motivated to teach equity-driven, justice-oriented content and skills. Additionally, tools for data visualizations are more often designed for STEM classes than for social studies classes. Our work attempts to address these challenges. In this paper, we share a few examples of technology and learning supports we have designed specifically to help social studies teachers implement equity-driven, justice-oriented data literacy and discuss what we have learned from implementation.

## Social Studies as a Venue for Teaching Equity-Driven, Justice-Oriented Data Literacy

In its most basic form, data literacy is the ability to comprehend, analyze, interpret, evaluate, create, and argue with data and data visualizations (Carlson et al., 2011; D'Ignazio and Bhargava, 2015). It involves knowing what data mean, being able to draw accurate conclusions from data and data visualizations, and recognizing when data are being used or visualized in misleading or inappropriate ways. However, as interest in teaching data literacy in K-12 schools has increased, several scholars (e.g., D'ignazio and Klein, 2020; Irgens et al., 2020; Lee, Wilkerson, and Lanouette, 2021) have called attention to the importance of critical and humanistic stances toward data and data visualizations that consider the personal, cultural, and sociopolitical factors involved with data collection, analysis, interpretation, and representation. They argue that from generation and collection through representation and visualization, data are never neutral and objective and that data can be manipulated to mislead or deceive, sometimes reinforcing or promoting oppressive stereotypes or structures (Gillborn, Warmington, and Demack, 2018; Hullman and Diakopoulos, 2011; Irgens et al., 2020; Philip, Olivares-Pasillas, and Rocha, 2016). Therefore, students should not just comprehend data and data visualizations, but should also engage in domain-specific, open discourses and critiques around data's usage and representation and the ways in which data can produce and reproduce injustice. They need to ask themselves who collected these data and for what purpose, from whom were the data collected, and who was left out.

Furthermore, data visualizations are narrative structures that involve a dynamic interplay between authorial intention and user interaction, and where ideological meanings are intentionally and unintentionally encoded by creators and "decoded" by end users (Philip et al., 2016; Philip and Rubel, 2019). Neither a direct correspondence between creator and end user, nor uniform interpretations among different end users, can ever be assumed (Philip et al., 2016; Philip and Rubel, 2019). Thus, teachers should consider students' personal connections with data and how their personal experiences will influence their interpretations (Johnson et al., 2021; Lee et al., 2021). Instruction that overemphasizes basic data literacy practices, without opportunities to discuss inequities in data practices and the ideological, historical, cultural, and sociopolitical dimensions of data production and usage, risks oversimplifying data, and data visualizations at best and reinforcing oppressive systems at worst (Philip and Rubel, 2019).

Unfortunately, a recent evaluation of curriculum frameworks inside and outside of STEM subjects indicates that teaching critical, justice-oriented data literacy skills is mostly absent from recommendations for data science instruction in schools (Drozda et al., 2022). Drozda et al. showed that although all learning frameworks that can guide data science instruction demonstrate the value and possible limitations of data, most do not consistently or deeply address the ethical and civic implications of data use in society. The *Launch Years Data Science Course Framework* from the University of Texas at Austin's Charles A. Dana Center is highlighted by Drozda et al. as a rare data science curriculum that considers data ethics, such as systemic bias in datasets. The framework states that students should "identify bias and sources of bias in data, and describe how bias in data impacts people and society". However, as of 2022, only 20 states have joined the Launch Years initiative, and we have yet to understand how influential the initiative has been on classroom practices and student learning. Actual implementation of a curriculum tends to vary dramatically from what its designers intended (Century, Rudnick, and Freeman, 2010), and the most successful implementation efforts involve a mutual adaptation process (Reiser et al., 2000) where designers, researchers, and teachers work together in partnership (Fishman et al., 2013). We are far too early in the development of data science curricula to see what outcomes they will achieve with students.

Drozda et al. (2022) considered the *K-12 CS Framework* (Yongpradit et al., 2016) in their review of data science implementations. The framework was developed in partnership with the Computer Science Teachers Association (CSTA), whose standards influence most of the state standards. Equity is a key concern in the framework, but with respect to computing overall. For example, the framework calls for students to consider who has access to computing, and how the needs of diverse end users are addressed in the design of computing systems. Data and Analysis is a core concept in the framework, which includes ideas like how data are stored and how algorithms process data, but not the kinds of critical questions that the Dana Center's curriculum raises.

Too often, equity is an afterthought in CS courses (Scott, 2020). While 90% of U.S. CS teachers feel that they have control over how they teach in their classroom, 60% do not teach about impacts of computing, and 58% do not cover ethics of computing (Koshy et al., 2022). Even if CS courses did address ethical issues around data collection and dissemination and promote justice-oriented data literacy, student access to such coursework is not equitable—only 5% of U.S. high school students take CS courses (https://advocacy.code.org/stateofcs). If our goal is to help students develop data literacy skills, CS classes are not the way to reach *many* students.

Social studies classes can be the avenue for addressing this problem. We offer four arguments for why student development of data literacy belongs to social studies.

- Data literacy helps students learn social studies.
- The subject of social studies has an equity- and social justice-oriented mission.
- A historical thinking lens changes how data literacy is taught.
- Everyday data literacy tends to be in social studies contexts.

First, data literacy fits the social studies curriculum and helps students learn social studies content. Social studies standards from all 50 states and the District of Columbia indicate that teachers in social studies courses are supposed to teach about data and data visualizations. The earliest explicit references to data visualizations begin in kindergarten for all but five U.S. states, where references begin shortly thereafter, in grade 1 or 2. Even where there is an absence of explicit references, standards documents frequently imply the use of data visualizations by recommending that students learn about movement, growth, or other changes over space and time (Shreiner, 2019).

Data and data visualizations are also prominent sources of information in social studies instructional resources. For example, students who are required to read from a textbook in their social studies classes will encounter a data visualization about every 13 pages in elementary school, every seven pages in middle school, and every four pages in high school. On average, 90% of data visualizations in social studies textbooks provide students with extensional information that is not found in the surrounding verbal text (Shreiner, 2018a). Moreover, resources from curricular websites that social studies teachers often rely upon, such as Stanford History Education Group, Library of Congress, Edsitement, and C3 Teachers, have data visualizations in slide presentations, readings, and document sets (Finholm and Shreiner, 2022). Teachers who even only occasionally use these resources are already teaching students with data visualizations.

There is also evidence to suggest data visualizations can enhance student learning in social studies. Reading data visualizations can improve students' overall comprehension and their quality of reasoning about social studies concepts like change over time and causation. Furthermore, data visualizations often contain background knowledge that students can later use to understand references to people, places, events, or documents (Norman, 2012; Roberts, Norman, and Cocco, 2015; Schnotz, 2002; Shreiner, 2019).

Second, teaching for equity and social justice is integral to what social studies teachers are expected and to do. The flagship organization for social studies teachers, the National Council for the Social Studies (NCSS; 2016), argues the mission of social studies teachers includes helping students "think critically", "make personal and civic decisions on information from multiple perspectives", and "develop a commitment to social responsibility, justice and action" (pp. 180–182). In line with this mission, NCSS's *College, Career, and Civil Life (C3) Framework for Social Studies State Standards*, a document that has guided the curricular efforts within the social studies profession since its publication (New et al., 2021), argues that social studies should equip students to address "societal problems" in part by emphasizing critical evaluation of evidence (National Council for the Social Studies, 2013, p. 6). Among the sources of evidence the *C3 Framework* includes in their recommendations are data and data visualizations such as maps and graphs, which student should find, evaluate, and use to build arguments and explanations (Drozda et al., 2022; Shreiner, 2019). Unfortunately, most states' social studies standards documents, including those that have been revised to meet the *C3 Framework*'s recommendations, fall short of providing explicit guidance for teaching critical and justice-oriented data literacy and not all social studies teachers are motivated to teach with such goals in mind (Peck and Herriot, 2014; Shreiner, 2019). However, there is still a constellation of guidance within the social studies teaching profession to orient teachers to teach critical, justice-oriented data literacy anyway, particularly if they are provided the tools to do so.

Third, social studies teachers have the unique capacity to teach data literacy in a way that other disciplinary teachers might not, such as teaching with a historical thinking lens. There is a decades-long movement in history to teach the habits of mind used by historians when they make sense of the past. These practices place considerable emphasis on the way students work with different forms of evidence, including data visualizations. Historical thinking entails an understanding that history is interpretation based upon remnants from the past and that interpretations are dependent on individual authors' biases and worldviews. Therefore, when approaching evidence, historically literate individuals must engage in a process called *sourcing*, whereby they identify the author and their biases and motivations, as well as take stock of where and when the document was produced (Leinhardt and Young, 1996; Rouet et al., 1997; Wineburg, 1991). They should also *contextualize* evidence, zooming out from the immediate context of the document and considering how concurrent events and trends might have influenced the content or intent of the evidence (Leinhardt and Young, 1996; van Drie and van Boxtel, 2008; Wineburg, 1991). In addition, historical thinking involves *corroboration* of evidence—checking for internal consistencies, comparing the evidence with other sources of information, and looking for discrepancies in information (Leinhardt and Young, 1996; Rouet et al., 1997; Wineburg, 1991).

These practices—sourcing, contextualizing, and corroborating— are critical when working with data and data visualizations. Social studies teachers are taught to implement these practices, likely more than STEM teachers. Social studies educators have lately emphasized the importance of properly evaluating social and political information on the internet, including statistical data and data visualizations (Breakstone et al., 2018; 2021; McGrew et al., 2018). Because social studies teachers already teach about the critical evaluation of historical evidence in their courses, they are likely better prepared than STEM teachers to help students apply these skills to data and data visualizations.

Finally, the everyday practice of data literacy needs to be framed more like social studies than any STEM subject, because how students approach data literacy and where they encounter the need for data literacy is far more likely to be related to social studies than STEM. Consider the work of Clegg et al. (2020) on the data literacy practices of Division 1 college athletes. They report that college athletes are engaged with data for many reasons: To assess their individual performance, to understand how and why their workouts changed, to connect workouts to performance goals, to consider how to change workouts in response to factors outside their control (e.g., classroom demands and injuries), and to use in communication with coaches and other colleagues. Some of this is quantitative reasoning (e.g., statistics), but some of it is about questioning the data (e.g., are calorie counts also considering nutrition?) and balancing the numbers with how the athlete "feels" and their sense of identity (Clegg et al., 2022). These are broader issues of data literacy, and not just statistics. While many of the data of college athletics might appear in science and mathematics classes, the framing needs to be in social studies terms. College athletes are rarely STEM majors (Svyantek et al., 2017). Among Division 1 college athletes, the most common majors are in the social studies cluster, with business second (Miller, 2021). Science and humanities are the least common clusters. If we want to reach the athletes, we have to reach them in social studies.

Furthermore, definitions of the social utility of data literacy (Pangrazio and Sefton-Green, 2020) and of data citizenship (Carmi et al., 2020) are not about the issues most commonly associated with data science in STEM subjects (Drozda et al., 2022). For example, Pangrazio and Sefton-Green are concerned with the impacts of data on a democracy – who is gathering data, who has access to it, and how citizens have insight into those processes. Those are explicitly the issues of civics and history classes in social studies curricula. Carmi et al. (2020) also share the concern about data literacy in a democracy, with a particular emphasis on who has control over the data. They emphasize the knowledge that citizens need to know about data in their daily lives, like the difference between dis-information, mis-information, and mal-information. They describe a framework of data citizenship with three areas (p. 10):

- **Data thinking**: Citizens' critical understanding of data.
- **Data doing**: Citizens' everyday engagements with data.
- **Data participation**: Citizens' proactive engagement with data and their networks of literacy, such as taking proactive steps to protect individual and collective privacy.

We ask the reader to engage in a thought experiment. In your daily life *outside* of work (since we assume that most readers are researchers or educators), where do you encounter data visualizations (including tables and infographics)? We expect that they are most likely in news sources. What are the subjects of those data visualizations? How is the visualization trying to influence you? The most important of these visualizations are likely conveying information about the state of the society or economy, and they are likely trying to change your behavior – either your everyday behavior (such as when you eat or commute) or in your role as a citizen (such as in the voting booth or to prompt you to write your governmental representative). Even when the topics of data visualizations in the news are scientific, such as climate change or infectious diseases, they are still presented to influence civic and political decision-making. These are social studies contexts. If we hope that we are preparing students to transfer their knowledge from the classroom into everyday life, social studies is where we might find the closest match to everyday use of data literacy.

## Supporting Social Studies Teachers in Teaching Data Literacy

There are significant challenges to incorporating data literacy into social studies classes. As mentioned earlier, all 50 of the United States include data literacy in their social studies standards (Shreiner, 2019). Yet, few social studies teachers actually incorporate data literacy into their classrooms (Shreiner and Dykes, 2020). The authors have been working together to support social studies teachers in incorporating data literacy for nearly five years. The rest of this section describes some of the lessons we have learned.

We frame these lessons with a positionality statement. The first author is a white, female history professor specializing in social studies education and a former social studies teacher. She teaches a class on data literacy in social studies to pre-service teachers. The second author is a white, male computing education researcher who is a senior member of his community. Our shared goal is to support social studies teachers in becoming *participants* in data literacy, in the data citizenship terms of (Carmi et al., 2020). We want teachers to understand how computing can be used to manipulate and visualize data. Our goal is for social studies teachers to be confident, informed, and capable users of computing for data literacy, as opposed to asking the teachers to become data scientists or software developers. Who we are and our goals biases how we see our results and influences the lessons learned that we present here.

From our previously published work (cited throughout the following paragraphs), we identify four key lessons for supporting social studies teachers in incorporating data literacy into their classroom practices and content.

**First, the existing data visualization tools that we have found are not well-designed for social studies teachers**. We started our work with participatory design sessions with pre-service social studies teachers. We provided scaffolding to help teachers to use several different programming languages for building data visualizations (e.g., JavaScript and Vega-Lite (Satyanarayan et al., 2017)) and direct manipulation data visualization tools (such as CODAP (Finzer and Damelin, 2016)). We then asked teachers what they liked and what they disliked about these tools, and what would be the perfect tool for their classrooms (Naimipour, Guzdial, and Shreiner, 2019b, 2020).

We continued these sessions with practicing teachers, moving to on-line sessions during the pandemic and continuing on-line today (Naimipour et al., 2021; Shreiner, Guzdial, and Naimipour, 2021). Our teachers are now supported with an Open Education Resource (OER) on data literacy for social studies which include examples of many tools for data visualization (Figure 7.1). Each linked tool is supported with a minimal manual (Anderson, Knussen, and Kibby, 1993; Carroll et al., 1987) that guides teachers successfully through an activity relevant to their classroom practice (Shreiner and Dykes, 2020). The OER includes modules on what is data literacy, why teach data literacy in social studies, what kinds of data visualizations will students encounter in social studies, and how should students analyze data and create and integrate data visualizations in social studies (Shreiner and Guzdial, 2022). The OER includes all the tools that we had used in our earlier sessions along with others that have been used in social studies classrooms.

All the tools that we introduced *worked* in the sense that teachers successfully built data visualizations. All the tools had features that the teachers found valuable. Those features that the teachers valued were carefully noted as inputs into our design process. However, none of them were tools that teachers were willing to incorporate into their practice.

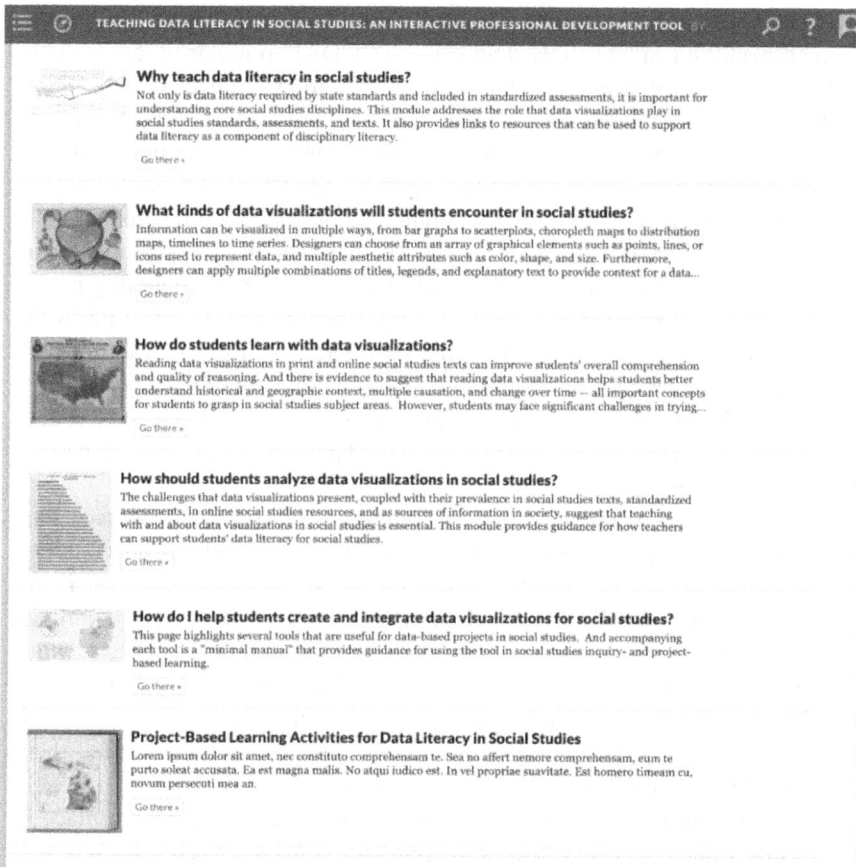

**FIGURE 7.1**   Open Education Resource (OER) for data literacy in social studies.

Let's consider two of these tools in a bit more detail to explain: Timeline JS and CODAP. The most used data visualizations in history classes are timelines and maps. Timeline JS from the Knight Lab at Northwestern[1] produces high-quality timelines for the purposes of journalists. Users create a Google Spreadsheet with all the data needed for the timeline, then the spreadsheet is shared and its URL is entered into the Timeline JS website. The user is then provided with a URL that presents the timeline formatted for Web browsers. Many of our social studies teacher participants used Timeline JS successfully. Some of them built lesson plans around it. Most teachers found it forbidding. The most critical usability problem was that the spreadsheet had to be exactly formatted for Timeline JS to work. Teachers worried about deleting a row, or mistyping a word. If the spreadsheet did not work for Timeline JS, there was simply no output. No explanatory messages were displayed.

CODAP is a terrific tool that both authors have used in their own classrooms. A spreadsheet file can be simply dragged into CODAP to import it, and column headers can be dragged into graph or map objects to create data visualizations. Many of our teachers reported enjoying exploring the use of CODAP on their own. But when it came to adoption in their classrooms, our social studies teachers found CODAP to have low usability. One teacher estimated that it would take her about three hours of class time for her middle school students to be productive in CODAP, which was too high a cost for the benefit. Other teachers said that it was easy to create a graph, but they struggled to change the graph or to modify it.

The issue is not that the tools were not well-designed. They were not well-designed for the uses of our social studies teachers. Our teachers did not plan to work with data visualizations often enough to learn complicated procedures. Tools were often designed around paradigms of use (e.g., input spreadsheet data to get a timeline out) that are common in CS and perhaps across STEM classes, but are uncommon in social studies.

**Second, we are strong advocates for participatory design processes**. Participatory design was invented to support workers in Scandinavian countries who wanted a say in the design of tools that they would be using for many hours each day (Bødker, Kensing, and Simonsen, 2009; Druin, 1999; Hemmings et al., 2002). Increasingly, participatory design methods are being used to work with teachers, students, and designers in created learning environments (DiSalvo, 2016; DiSalvo et al., 2017; Könings, Seidel, and van Merriënboer, 2014; Wilkerson, 2017). We have used participatory design methods and find them invaluable for our purposes (Shreiner et al., 2021).

The critical problem is that data literacy does not currently have a place in most social studies teachers' practices (Shreiner and Dykes, 2021). They will easily dismiss a technology for data literacy that is not usable enough or not useful enough. Our challenge is to discern what they find attractive, interesting, and engaging enough that it is worth making the effort to incorporate into their classroom.

There are no landscape surveys to determine the needs of social studies teachers for incorporating data literacy. Further, it's not obvious what questions to ask on that survey. We find that teachers often surprise us, like when they told us what they liked about Vega-Lite, a visualization tool with an unusual textual syntax (Naimipour, Guzdial, and Shreiner, 2019a). We need to explore their preferences with concrete examples and personal experiences.

**Third, the data literacy tools that we create for social studies teachers have features different than most data visualization tools**. Data literacy tools are typically designed for professionals like journalists or data scientists, or they are designed for STEM classrooms. Social studies teachers are different from those professionals, and social studies classrooms are different from STEM classrooms.

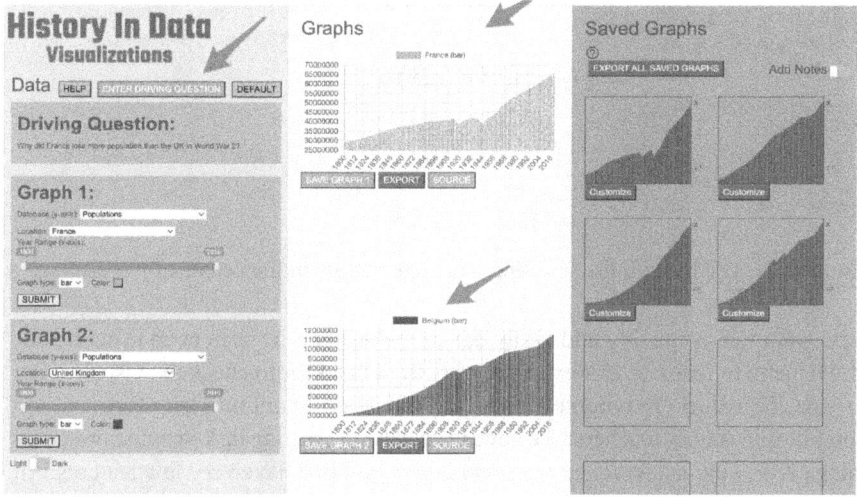

**FIGURE 7.2**    DV4L with arrows pointing to the driving question and to the two visualizations appearing in the center.

We have been developing a data visualization tool for history classes called DV4L (Figure 7.2) whose design is informed by our participatory design process (Naimipour et al., 2019a, 2020; Naimipour, Shreiner, and Guzdial, 2020). While the tool has gone through many iterations of development at this point, a constant feature is that two data visualizations are always presented at once. This feature is informed by history education practice that a historical problem arises when two accounts, visualizations, or data contradict (Bain, 2005; 2009). Therefore, we always present two visualizations so that two datasets (e.g., two nations historical data for the same variable) can be easily compared. No visualization tool that we have found so far makes it easy to create and compare two data visualizations. That is not a common practice in STEM classes, but it's part of history practice.

DV4L prominently asks the student to "Enter a Driving Question" (upper left of the display in Figure 7.2), and then displays that driving question as they work. Driving questions are a key part of inquiry, problem, and project-based learning (Blumenfeld et al., 1994; Krajcik et al., 1998; Krajcik and Blumenfeld, 2006; Quintana, Krajcik, and Soloway, 2002). Our social studies teachers asked that our data literacy tool be organized around a driving question, because it is key to how they teach. DV4L may be the only data literacy tool for schools that prompts for and displays a driving question, because it is built for social studies teachers and that was important for them.

There are times when a programming language is useful for data visualizations in a social studies classroom. In particular, teachers have told us that having

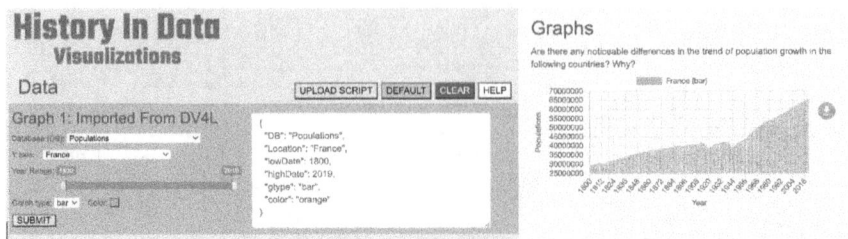

**FIGURE 7.3**   DV4L with the scripting language visible in the center of the display.

a program that is quickly and easily edited makes it easier than even using a pull-down menu to focus on (for example) particular dates, or to check the visualization for specific values. In our experience, traditional programming languages are too complicated for our teachers. We have instead been creating task-specific programming (TSP) languages, called *teaspoon languages*, that have very few features. All the features in our DV4L teaspoon language (Figure 7.3) are specific to creating a data visualization (Guzdial and Naimipour, 2019; Guzdial and Shreiner, 2021).

We are working on new tools that meet needs that social studies teachers tell us about. Most visualization tools (including our own) present a graph with all its pieces: X and Y axes, data points, and legend. We know that many students do not interpret and comprehend all of those visual elements when they first see a visualization, but the more elements they do interpret, the more that they learn from the visualization (Shreiner, 2018b). We are working to help teachers incorporate *slow reveal graphs* into their practice (Taurence, Shreiner, and Dykes, 2022). A slow reveal graph introduces a graph element-by-element to help students to see all the visual elements without overwhelming them. We are working on new tools to support teachers in defining a *slow analysis* of an existing visualization, by masking all but one element at a time with a prompt to direct student attention. Again, these are tools that we are developing because of the needs of practicing teachers.

**Fourth, resources and tools alone are insufficient for adoption**. The OER for data literacy in social studies education is a tremendous resource for teachers. They tell us that in every session that we run with them. But it has not led to significant adoption of data literacy in our teachers' social studies classrooms. The barriers are significant, and they cannot all be addressed with the resources and tools that we can provide.

When we started our work, we drew upon the Technology Acceptance Model (TAM) (Aldunate and Nussbaum, 2013; Lee, Kozar, and Larsen, 2003; Teo, 2009) which generally says that adoption of a technology occurs when it is perceived as highly usable and highly useful. Based on what our teachers tell us at the end of our session, our resources and tools have met those criteria, but we still do not see much adoption. We realize that there are issues beyond TAM's usability and usefulness (Shreiner and Guzdial, 2022).

Teachers must feel confident and informed. They need to feel supported within their school context. If no other social studies teacher in a school is teaching data literacy, it's hard for even the well-informed and confident new teacher to start. Providing tools and resources are a critical first step. But there are social and personal issues that also must be met to gain adoption.

In this chapter, we argue for the importance of data literacy in the social studies classroom. Social studies is a critical place to teach equity-focused data literacy because these are the right teachers in the right context for impact in students' everyday lives as data literate citizens. However, we still are learning how to support data literacy in the social studies classroom. It's a critical need for just and equitable data science education.

## Acknowledgments

We thank Bahare Naimipour and our undergraduate research team for their contributions to this project. Some of this material is based upon work supported by the National Science Foundation under Grant No. 2030919. Any opinions, findings, and conclusions or recommendations expressed in this material are those of the author(s) and do not necessarily reflect the views of the National Science Foundation.

## Note

1 https://timeline.knightlab.com/

## References

Aldunate, R., & Nussbaum, M. (2013). Teacher adoption of technology. *Computers in Human Behavior*, *29*(3), 519–524.

Anderson, A., Knussen, C. L., & Kibby, M. R. (1993). Teaching teachers to use Hyper-Card: A minimal manual approach. *British Journal of Educational Technology*, *24*(2), 92–101.

Bain, R. B. (2005). They thought the world was flat? Applying the principles of how people learn in teaching high school history. In S. Donovan, J. Bransford, & A. T. R. f. T. National Research Council Committee on How People Learn (Eds.), *How students learn history in the classroom* (pp. 179–214). National Academies Press. http://www.loc.gov/catdir/toc/ecip053/2004026246.htmlhttp://firstsearch.oclc.org/WebZ/DCARead?standardNoType=1&standardNo=0309074339:srcdbname=worldcat:fromExternal=true&sessionid=0

Bain, R. B. (2009). Into the breach: Using research and theory to shape history instruction. *Journal of Education*, *189*(1–2), 159–167.

Blumenfeld, P. C., Krajcik, J., Marx, R. W., & Soloway, E. (1994). Lessons learned: A collaborative model for helping teachers learn project-based instruction. *Elementary School Journal*, *94*(5), 539–551.

Bødker, K., Kensing, F., & Simonsen, J. (2009). *Participatory IT design: Designing for business and workplace realities*. MIT press.

Breakstone, J., McGrew, S., Smith, M., Ortega, T., & Wineburg, S. (2018). Teaching students to navigate the online landscape. *Social Education, 82*(4), 219–221.

Breakstone, J., Smith, M., Wineburg, S., Rapaport, A., Carle, J., Garland, M., & Saavedra, A. (2021). Students' civic online reasoning: A national portrait. *Educational Researcher, 50*(8), 505–515.

Carlson, J., Fosmire, M., Miller, C., & Nelson, M. S. (2011). Determining data information literacy needs: A study of students and research faculty. *Portal: Libraries and the Academy, 11*(2), 629–657.

Carmi, E., Yates, S. J., Lockley, E., & Pawluczuk, A. (2020). Data citizenship: Rethinking data literacy in the age of disinformation, misinformation, and malinformation. *Internet Policy Review, 9*(2), 1–22.

Carroll, J. M., Smith-Kerker, P. L., Ford, J. R., & Mazur-Rimetz, S. A. (1987). The minimal manual. *Human-Computer Interaction, 3*(2), 123–153.

Century, J., Rudnick, M., & Freeman, C. (2010). A framework for measuring fidelity of implementation: A foundation for shared language and accumulation of knowledge. *American Journal of Evaluation, 31*(2), 199–218.

Clegg, T., Greene, D. M., Beard, N., & Brunson, J. (2020). Data everyday: Data literacy practices in a Division I college sports context. In *Proceedings of the 2020 CHI conference on human factors in computing systems*.

Clegg, T. L., Cleveland, K., Weight, E., Greene, D., & Elmqvist, N. (2022). Data everyday as community-driven science: Athletes' critical data literacy practices in collegiate sports contexts. *Journal of Research in Science Teaching, 60*(8), 1786–1816. https://doi.org/10.1002/tea.21842

D'ignazio, C., & Klein, L. F. (2020). *Data feminism*. MIT press.

D'Ignazio, C., & Bhargava, R. (2015). Approaches to building big data literacy. In *Proceedings of the Bloomberg data for good exchange conference*.

DiSalvo, B. (2016). Participatory design through a learning science lens. In *ACM Proceedings of the 2016 CHI conference on human factors in computing systems*.

DiSalvo, B., Yip, J., Bonsignore, E., & DiSalvo, C. (2017). *Participatory design for learning*. Taylor & Francis.

Drozda, Z., Johnstone, D., & Van Horne, B. (2022). *Previewing the national landscape of K-12 data science implementation*. E. National Academies of Sciences, and Mathematics. https://www.nationalacademies.org/event/09-13-2022/docs/D688ED916E82498 DA0E2171A109936D679FD5DE26556

Druin, A. (1999). Cooperative Inquiry: Developing New Technologies for Children with Children. In *CHI '99 Proceedings of the SIGCHI Conference on Human Factors in Computing Systems*.

Finholm, C. E., & Shreiner, T. L. (2022). A lesson in missed opportunities: Examining the use of data visualizations in online history lessons. *Social Studies Research and Practice, 17*(2), 155–166.

Finzer, W., & Damelin, D. (2016). Design perspective on the Common Online Data Analysis Platform (CODAP). In *American Educational Research Association (AERA) conference*.

Fishman, B. J., Penuel, W. R., Allen, A.-R., Cheng, B. H., & Sabelli, N. (2013). Design-based implementation research: An emerging model for transforming the relationship of research and practice. *National Society for the Study of Education, 112*(2), 136–156.

Gillborn, D., Warmington, P., & Demack, S. (2018). QuantCrit: Education, policy, 'Big Data' and principles for a critical race theory of statistics. *Race ethnicity and education, 21*(2), 158–179.

Guzdial, M., & Naimipour, B. (2019). Task-specific programming languages for promoting computing integration: A precalculus example. In *Koli Calling '19 Proceedings of the 19th Koli Calling International Conference on Computing Education Research.*

Guzdial, M., & Shreiner, T. (2021). Integrating computing through task-specific programming for disciplinary relevance: Considerations and examples. *Computational thinking in education* (pp. 172–190). Routledge.

Hemmings, T., Crabtree, A., Rodden, T., Clarke, K., & Rouncefield, M. (2002). Probing the probes. In *Proceedings of the participatory design conference.*

Hullman, J., & Diakopoulos, N. (2011). Visualization rhetoric: Framing effects in narrative visualization. *IEEE Transactions on Visualization and Computer Graphics, 17*(12), 2231–2240.

Irgens, G. A., Knight, S., Wise, A. F., Philip, T. M., Olivares, M. C., van Wart, S., Vakil, S., Marshall, J., Parikh, T., Lopez, M. L., Wilkerson, M. H., Gutiérrez, K., Jiang, S., & Kahn, J. B. (2020). Data literacies and social justice: Exploring critical data literacies through sociocultural perspectives. *14th International conference of the learning sciences: The interdisciplinarity of the learning sciences, ICLS 2020.*

Johnson, B., Rydal Shapiro, B., DiSalvo, B., Rothschild, A., & DiSalvo, C. (2021). Exploring approaches to data literacy through a critical race theory perspective. In *Proceedings of the 2021 CHI conference on human factors in computing systems.*

Kahne, J., & Bowyer, B. (2017). Educating for democracy in a partisan age: Confronting the challenges of motivated reasoning and misinformation. *American Educational Research Journal, 54*(1), 3–34.

Könings, K. D., Seidel, T., & van Merriënboer, J. J. G. (2014). Participatory design of learning environments: Integrating perspectives of students, teachers, and designers. *Instructional Science, 42*(1), 1–9.

Konold, C., Higgins, T., Russell, S. J., & Khalil, K. (2015). Data seen through different lenses. *Educational Studies in Mathematics, 88*(3), 305–325.

Koshy, S., Twarek, B., Bashir, D., Glass, S., Goins, R., Novohatski, L. C., & Scott, A. (2022). *Moving Towards a Vision of Equitable Computer Science: Results of a Landscape Survey of PreK-12 CS Teachers in the United States.* https://landscape.csteachers.org/

Krajcik, J., Blumenfeld, P. C., Marx, R. W., Bass, K. M., Fredricks, J., & Soloway, E. (1998). Inquiry in project-based science classrooms: Initial attempts by middle school students. *Journal of the Learning Sciences, 7*(3,4), 313–350.

Krajcik, J. S., & Blumenfeld, P. C. (2006). Project-based learning. In R. K. Sawyer (Ed.), *The Cambridge handbook of the learning sciences.* Cambridge University Press.

LaMar, T., & Boaler, J. (2021). *The importance and emergence of K-12 data science. Phi Delta Kappan, Issue.*

Lee, Y., Kozar, K. A., & Larsen, K. R. T. (2003). The technology acceptance model: Past, present, and future. *Communications of the Association for Information Systems, 12*(1), 50.

Lee, V. R., Wilkerson, M. H., & Lanouette, K. (2021). A call for a humanistic stance toward K–12 data science education. *Educational Researcher, 50*(9), 664–672.

Leinhardt, G., & Young, K. M. (1996). Two texts, three readers: Distance and expertise in reading history. *Cognition and Instruction, 14*(4), 441–486.

Levitt, S. D. (2022). Data science is the future. Let's start teaching it. *Education Week.* https://www.edweek.org/teaching-learning/opinion-data-science-is-the-future-lets-start-teaching-it/2022/01

McGrew, S., Breakstone, J., Ortega, T., Smith, M., & Wineburg, S. (2018). Can students evaluate online sources? Learning from assessments of civic online reasoning. *Theory and Research in Social Education, 46*(2), 165–193.

Miller, S. (2022). An analysis of the rate of academic clustering and the types of majors chosen by divisions I, II, and III intercollegiate athletes. *Journal for the Study of Sports and Athletes in Education, 16*(2), 97–113.

Naimipour, B., Guzdial, M., & Shreiner, T. (2019a). Helping social studies teachers to design learning experiences around data: Participatory design for new teacher-centric programming languages. *ICER '19* Proceedings of the 2019 ACM conference on international computing education research.

Naimipour, B., Guzdial, M., & Shreiner, T. (2019b). *Helping social studies teachers to design learning experiences around data: Participatory design for new teacher-centric programming languages.* Proceedings of the 2019 ACM conference on international computing education research. http://doi.acm.org/10.1145/3291279.3341211

Naimipour, B., Guzdial, M., & Shreiner, T. (2020). *Engaging pre-service teachers in front-end design: Developing technology for a social studies classroom.* In *2020 IEEE Frontiers in Education Conference (FIE).*

Naimipour, B., Guzdial, M., Shreiner, T., & Spencer, I. (2021, March). From guided exploration to possible adoption: patterns of pre-service social studies teacher engagement with programming and non-programming based learning technology tools. In *Proceedings of the 2021 Society for Information Technology and Teacher Education (SITE) Interncational Conference* (pp. 1508–1513). Association for the Advancement of Computing in Education (AACE).

Naimipour, B., Shreiner, T. L., & Guzdial, M. (2020). Engaging Teachers in Front-End Design: Developing Technology for a Social Studies Classroom. Proceedings of the ASEE/IEEE 2020 Frontiers in Education Conference.

National Council for the Social Studies. (2013). *The College, Career, and Civic Life (C3) Framework for Social Studies State Standards: Guidance for Enhancing the Rigor of K-12 Civics, Economics, Geography, and History.* NCSS.

National Council for the Social Studies (2016). A vision of powerful teaching and learning in the social studies. *Social Education, 80*(3), 180–182.

National Council for the Social Studies. (2021). *About NCSS.* National Council for the Social Studies. http://www.socialstudies.org/about/

New, R., Swan, K., Lee, J., & Grant, S. (2021). The state of social studies standards: What is the impact of the C3 framework? *Social Education, 85*(4), 239–246.

Norman, R. R. (2012). Reading The graphics: What is The relationship between graphical Reading processes and student comprehension? *Reading and Writing, 25*(3), 739–774.

Pangrazio, L., & Sefton-Green, J. (2020). The social utility of 'data literacy. *Learning, Media and Technology, 45*(2), 208–220.

Peck, C. L., & Herriot, L. (2014). Teachers' beliefs about social studies. *International Handbook of Research on Teachers' Beliefs* (pp. 387–402) Routledge.

Philip, T. M., Olivares-Pasillas, M. C., & Rocha, J. (2016). Becoming racially literate about data and data-literate about race: Data visualizations in the classroom as site of racial-ideological micro-contestations. *Cognition and Instruction, 34*(4), 361–388. https://doi.org/10.1080/07370008.2016.1210418

Philip, T. M., & Rubel, L. (2019). Classrooms as laboratories of democracy for social transformation: The role of data literacy. In *Shifting contexts, stable core: Advancing quantitative literacy in higher education.* Mathematical Association of America.

Quintana, C., Krajcik, J., & Soloway, E. (2002). A Case Study to Distill Structural Scaffolding Guidelines for Scaffolded Software Environments. *CHI '02: Proceedings of the SIGCHI conference on Human factors in computing systems.*

Reiser, B. J., Spillane, J. P., Steinmuller, F., Sorsa, D., Carney, K., & Kyza, E. (2000). Investigating the Mutual Adaptation Process in Teachers' Design of Technology-Infused Curricula. In B. Fishman & S. O'Connor-Divelbiss (Eds.), *Fourth International Conference of the Learning Sciences* (pp. 342–349). Erlbaum.

Roberts, K. L., Norman, R. R., & Cocco, J. (2015). Relationship between graphical device comprehension and overall text comprehension for third-grade children. *Reading Psychology, 36,* 389–420. https://doi.org/10.1080/02702711.2013.865693

Rouet, J.-F., Favart, M., Britt, M. A., & Perfetti, C. A. (1997). Studying and using multiple documents in history: Effects of discipline expertise. *Cognition and Instruction, 15*(1), 85–106.

Satyanarayan, A., Moritz, D., Wongsuphasawat, K., & Heer, J. (2017). Vega-lite: A grammar of interactive graphics. *IEEE Transactions on Visualization and Computer Graphics, 23*(1), 341–350. https://doi.org/10.1109/TVCG.2016.2599030

Schnotz, W. (2002). Towards an integrated view of learning from text and visual displays. *Educational Psychology Review, 14*(1), 101–120.

Scott, A. (2020, January 10). Addressing equity in CS curriculum with Kapor Center. *Code with Google.* https://blog.google/outreach-initiatives/code-with-google/curriculum-equity/

Shreiner, T. L. (2018a). Data literacy for social studies: Examining the role of data visualizations in k-12 textbooks. *Theory and Research in Social Education, 46,* 194–231. https://doi.org/10.1080/00933104.2017.1400483

Shreiner, T. L. (2018b). Students' use of data visualizations in historical reasoning: A think-aloud investigation with elementary, middle, and high school students. *The Journal of Social Studies Research,* 43(4). https://doi.org/10.1016/j.jssr.2018.11.001

Shreiner, T. L. (2019). Students' use of data visualizations in historical reasoning: A think-aloud investigation with elementary, middle, and high school students. *The Journal of Social Studies Research, 43*(4), 389–404.

Shreiner, T. L., & Dykes, B. (2020). Teaching Data Literacy for Social Studies: Teacher Practices, Beliefs, and Knowledge. In *The American educational research association annual meeting.*

Shreiner, T. L., & Dykes, B. M. (2021). Visualizing the teaching of data visualizations in social studies: A study of Teachers' data literacy practices, beliefs, and knowledge. *Theory and Research in Social Education, 49*(2), 262–306.

Shreiner, T. L., & Guzdial, M. (2022). The information won't just sink in: Helping teachers provide technology-assisted data literacy instruction in social studies. *British Journal of Educational Technology, 53*(5), 1134–1158.

Shreiner, T. L., Guzdial, M., & Naimipour, B. (2021). *Using Participatory Design Research to Support the Teaching and Learning of Data Literacy in Social Studies.*

In *CUFA, the College and University Faculty Assembly 2021 of the National Council of the Social Studies*.

Svyantek, D. J., Connelly, B., O'Neill, S., Boudreaux, M., Struempler, B., & Teeter, L. (2017). Academic clustering among college athletes. *Sports and Understanding Organizations*, 283–298.

Taurence, K., Shreiner, T., & Dykes, B. (2022). Revealing the Power of Data Visualizations in Social Studies through Slow Reveal Graphs. *Statistics Teacher*(Spring). https://www.statisticsteacher.org/2022/03/23/slowrevealgraphs/

Teo, T. (2009). Modelling technology acceptance in education: A study of pre-service teachers. *Computers\& Education*, *52*(2), 302–312.

van Drie, J., & van Boxtel, C. (2008). Historical reasoning: Towards a framework for analyzing students' reasoning about the past. *Educational Psychology Research in Social Education*, *20*(2), 87–110. https://doi.org/10.1007/s10648-007-9056-1

Weiland, T., & Engledowl, C. (2022). Transforming curriculum and building capacity in K–12 data science education. *Harvard Data Science Review*, *4*(4).

Wilkerson, M. H. (2017). Teachers, students, and after-school professionals as designers of digital tools for learning. In *Participatory design for learning: Perspectives from research and practice*. Routledge.

Wineburg, S. (1991). On the reading of historical texts: Notes on the breach between school and academy. *American Educational Research Journal*, *28*(3), 495–519.

Yongpradit, P., Hendrickson, K., & Phillips, R. (2016). *K-12 CS framework*. https://k12cs.org/

# 8

# THE UTILITY OF DESIGNING DATA SCIENCE EDUCATION PROGRAMS FROM A FRAMEWORK OF IDENTITY

*June Ahn, Seth Van Doren, Jessica Cai, Ha Nguyen, Fernando Rodriguez, Christopher Martinez, and Jenny Han*

## Introduction

A fundamental conundrum in data science education is understanding how to foster equitable outcomes in the field. Data has become easier to collect, store, connect, aggregate, analyze, and integrate into a wide variety of professions or occupational pathways (Persaud, 2021). In this new world, data is a vitally important tool and currency. Immense power and opportunity are afforded to those who are literate with data or can work with data in increasingly more advanced and ethical ways (Lang et al., 2017; Siemens and Baker, 2012). The demand for data scientists continues to grow with diverse new career opportunities and roles proliferating rapidly (Davenport and Patil, 2022). However, this demand is likely to be met in inequitable ways. If history offers any lessons, students from more privileged backgrounds will likely be better positioned to take advantage of these emerging opportunities. For example, in the field of computer science (CS), gaps in the representation of women and racial minority groups in software programming careers have widened or remained, even with an overall increase in the diversity of students who obtained CS degrees (John and Carnoy, 2019). Such findings highlight how merely gaining training or educational credentials in a field such as CS does not adequately explain why individuals decide not to pursue futures in those fields or are systematically excluded in some way. How might we explore this conundrum, defined as the gap in opportunity versus equitable inclusion into the field?

In the following chapter, we outline an analytical lens that focuses on exploring how learners can come to identify with the field of data science, and in particular the process through which learners specifically might develop a

DOI: 10.4324/9781003364634-8

deeper identity to **become a data scientist** as they progress through their lives. This focus on identity development complements other potential ways to define an equitable outcome. For example, research in data science education often focuses on core knowledge and skills that are part of working with data (Kim, 2016; Persaud, 2021), expanding what we mean as being literate with data to move beyond just formal skills to also value personally relevant ways of engaging with data (Clegg et al., 2020; 2022), or figuring out ways to better integrate data work with social justice or humanistic experiences to hopefully connect with more learners (Calabrese Barton et al., 2021; Lee, Wilkerson, and Lanouette, 2021). We categorize this focus as attending to instructional concerns or identifying the diverse opportunities through which learners could engage with data practices or skills. Conversely, others might be focused on representation or general inclusion, which shifts attention to ensuring that diverse learners are represented in data science programs, classrooms, or jobs (Li and Koedel, 2017). This framing of equitable outcomes prioritizes expanding access to programs or jobs, but theoretically obscures what experiences might foster or lead to equitable representation.

In the following chapter, we outline a framework for how a learner's emerging occupational identity (Callahan et al., 2019) as a data scientist could be understood through five components: (1) one's self-positioning, (2) competency beliefs, (3) social capital, (4) structural opportunity, and (5) navigational understanding. Then we shed light on the utility of this identity framework by sharing case vignettes of undergraduate students, who identify with traditionally underrepresented groups, and who are in a university program that is designed to foster pathways into data science careers. The unique aspect of our case study work is that these students are a self-selected population who are already interested in data science as a field and career pathway. This participant sample is a limitation in the sense that we do not draw generalized claims from our research. However, this purposeful sample is theoretically interesting because we can observe the variability of experiences that students have, even when they share an already existing interest in the field.

Within this seemingly ideal student population (who one might expect are of the highest probability to pursue data science), we show how an identity lens helps reveal the complex interplay of social, economic, and political factors that each individual learner grapples with as they both experience the educational program and develop their own evolving decisions about whether they see themselves as data scientists. From these vignettes we illuminate how the *design of educational programs or pathway experiences* might attend to these complex factors, complement both instructional and access approaches to equity, and ultimately inform new research hypotheses about how we might promote data science experiences that attend to the social, economic, and political realities of students.

### Occupational Identity as a Focal Lens for Data Science Education

A lens of occupational identity as a core construct of interest (Callahan et al., 2019) attunes research directions on some core boundaries. First, we see learners' decisions to pursue and persist – long term – on careers in data science as a focal outcome (versus other valid outcomes such as learning technical data science skills etc.) and this focus often involves different experiences for learners than just those in formal classrooms (Calabrese Barton et al., 2013; Carlone and Johnson, 2007; Polman and Miller, 2010). Second, examining identity necessitates defining what aspects make up one's decision to "be a certain type of person" (Sfard and Prusak, 2005) and how learning environments support or hinder these complex decisions (Ahn, 2019). Identity can be defined in a myriad of ways and the construct is debated and complex. For example, sometimes identity is described as a process of learning to be a part of an established group, such as in a communities of practice framework (Lave and Wenger, 1991). Other times, identity is described as the narratives one might tell about oneself (Sfard and Prusak, 2005), a sense of belonging to a group (Jones, Tendhar, and Paretti, 2016), the ownership a learner might feel (Yip et al., 2014), or perhaps one's imagination of themselves (Holland et al., 2001).

Recognizing the complexity and different layers of identity, we focus our framework on five concepts that are particularly salient for the development of occupational identity, where a learner may develop a deeper sense that they are – for example, a data science person who pursues this identity as a career path. Below, we briefly describe the components that might explain a learner's occupational identity and explain how we conceptualize the components in the context of data science education.

### *Self-positioning: Imagining My Possible Self in a Field*

Learners iteratively evaluate and define possible selves—defined as "cognitive manifestations of goals, aspirations, values, and fears" (Shepard and Marshall, 1999)—in relation to specific situations or goals (Oyserman and Markus, 1990). For example, a learner may position their values, goals, and aspirations to align themselves with possible selves in different careers. This process can motivate them to pursue relevant, career-advancing opportunities (Brown et al., 2020; Markus and Nurins, 1986). In the context of data science education, we might ask learners to reflect on aspects of a future career (e.g., as a data scientist) that are appealing or unappealing to them, and emotions and questions that they have surrounding this career field. This reflection then reveals underlying values around the career and helps them reflect on potential alignment between their goals, aspirations, and values and how they imagine possible selves in the future.

### Competency Beliefs: Perceptions of One's Knowledge and Skills

We draw from social cognitive theory (Bandura, 1997; 2001) to capture learners' self-perceptions of their knowledge and skills in the data science field. Competency beliefs can be defined as the extent to which one perceives their ability within a domain to achieve the desired outcomes. One's competency beliefs can influence their behaviors, affects, and motivations, and sustain their commitment to pursuing career-advancing opportunities (Carpi et al., 2017; Eccles and Wigfield, 2002). Treating competency beliefs as an important component of identity development, we might ask learners to identify areas of knowledge and skills that they perceive needing to know to pursue careers as a data scientist, and their perceived competence in these areas. Supporting students' competency beliefs can lead to gains in how students see themselves fitting into science disciplines, promote their confidence, and enhance motivation to learn (Lopatto, 2007; Ryder, Leach, and Driver, 1999).

### Social Capital: Relationships that One Perceives as Beneficial

Learners pursue goals with the support of social capital—such as a network of families, peers, mentors, and other resources (Bourdieu, 2011). Putnam (2000) describes social capital as bonding and bridging social capital, both of which are necessary to support learners' identity construction. Bonding social capital describes deep, close ties that provide emotional and personal support. These relationships might serve as models that influence how learners self-position themselves and pursue opportunities. Meanwhile, bridging social capital represents less close, but diverse relationships that learners can build on when seeking information and expanded opportunities. Both bonding and bridging social capital can serve as embedded assets, as learners decide on and persist in a career field (Martin et al., 2020). Inviting learners to reflect on (1) their close relationships and (2) mentors in the desired career fields can help them articulate the robust social, emotional, and informational resources that they already possess. It also provides the opportunity to identify gaps in one's social network and broker new relationships, particularly when learners first enter a career field.

### Structural Opportunity: Access to Experiences that Learners Are Afforded

We further focus on structural opportunities, defined as activities relevant to the pursuit of the desired careers (Azevedo, 2011; Bohnert, Fredricks, and Randall, 2010; McDonnell, 1995). Not all students have access to activities and experiences traditionally valued in data science education, such as coming in with mathematics, statistics, and CS backgrounds (National

Academies of Science, Engineering, and Medicine, 2018). At the same time, students might already engage in data work through informal ways, such as reading, interpreting, and analyzing their own fitness data, or relating data patterns to personal and community experiences (Calabrese Barton et al., 2021; Clegg et al., 2020). It is thus critical to highlight how their everyday experiences and activities are contributing to future careers (Takeuchi, Vaala, and Ahn, 2019). In helping learners to consider structural opportunities that support future careers as data scientists, we ask them to reflect on (1) the opportunities to practice skills or qualities that are connected to the specific career, and (2) everyday experiences that can also be relevant to the career. These questions help us to understand opportunities that learners already have, while also elevating the everyday, informal experiences they already engage in.

### *Navigational Understanding: Knowledge to Take Advantage of Opportunities and Engage with Organizations or Institutions*

Learners also need to navigate multiple layers of requirements in their pathways to a career, from academic (e.g., knowing how to get into college, completing major-related prerequisites, etc.) to career-specific milestones (e.g., obtaining an internship, getting a certain work certification, etc.). Learners need to identify the steps they need to take to progress toward a career, reflect on existing experiences, values, and interests, and plan logistically for how they might pursue next steps. One may have a broad concept of their future career goals, but identifying concrete next steps that map onto specific time frames can be challenging for many learners (Brown et al., 2020). Thus, we invite learners to concretely explore (1) the challenges, requirements, and opportunities that exist to get them to their desired careers as data scientists, and (2) the specific steps they are taking within the short and long term. The goal of this exercise is for learners to continually reflect on their navigation and develop a coherent path of action over time, rather than dictating a single correct path.

### Observing This Identity Framework in Action: Case Studies of Learner Experiences

In our ongoing work, we are developing case study accounts of how undergraduate students develop their occupational identities as they participate in a specialized training and mentorship program in data science. The data science program is housed in a research-intensive, large public university that serves a large population of low-income students, first generation college students, and is a designated Hispanic Serving Institution and Asian American, Native American and Pacific Islander Serving Institution. The program is selective and competitive, where each fall undergraduate students across the whole campus are invited to apply to participate. A cohort of 15 students are chosen each year to participate in a year-long fellowship that lasts from January-December of the student's 3rd and 4th year of their undergraduate studies. Thus, students typically start the program at the end of their junior year and complete the fellowship as they are applying to data science graduate programs in the beginning of their senior year of undergraduate study.

The program is designed to support students who voice an interest in pursuing a career in learning analytics or data science and attending graduate school. Priority is given to students who identify with traditionally underrepresented populations, including students from minority racial/ethnic backgrounds (Latine, Black, Indigenous, Pacific Islander), first-generation college students, disabled students, and students from economically disadvantaged backgrounds. During the year-long program, fellows propose and undertake a data science research project under the guidance of a faculty mentor. Fellows are also encouraged to participate in the faculty mentor's research lab or group, to gain experience in the cultural and social aspects of working in university research settings. Fellows receive additional training through bi-weekly professional development sessions where they participate in an R studio workshop and a research seminar, where they learn skills in using R and conducting research. During the summer months, fellows participate in their research labs full time and begin preparing for graduate school applications. The program financially supports fellows through monthly stipends, room and board during the summer session, and by paying for GRE test preparation and graduate school applications.

### Participants and Initial Case Analyses

Of the 14 fellows in the program – who also participate in our ongoing research study – 13 identify as women and one as a man. In terms of racial/ethnic identity: four fellows identify as Hispanic/Latine, four as Southeast Asian, five as East Asian, and one as White. Six of the fellows shared that they are first-generation college students, and one identifies as being disabled. To gain deep understanding of the five components of occupational identity trajectories of our fellows, and how the different components shift and transform over time, we observe the seminars they attend, and schedule a series of interviews with our fellows throughout their fellowship year. During the seminars we produce field notes and analytical memos, taking special attention to the interactions and methods of facilitation at play. At the time of writing this chapter, 18 weeks into the program, we have conducted two interviews with each fellow, one just after fellows have been onboarded and integrated into their research labs (three to four weeks into the program) and the second after fellows have identified a potential individual project to pursue (10–11 weeks into the program). Interviews follow a semi-structured interview protocol with items that help us understand each fellow's beliefs and perceptions about their self-positioning, competency beliefs, social capital perceptions, structural opportunities, and navigational knowledge. Three more interviews are scheduled for the remainder of the program. Across these interviews, our research aim is to trace the evolution of how the data science fellows put together these different identity facets, in their own narratives.

In the following section, we shed light on the affordances of tracing the students' learning through the lens of these identity facets. Specifically, we show how the impactful moments that different fellows remember, perceive, and share with us throughout the program can be observed as a complex intertwining of the five elements of our framework. This understanding then helps us as data science education researchers – and educators or mentors – to reflect on how different aspects of this university program interact with each individual student in variable ways. To illustrate the utility of this analysis, we offer two example vignettes in this chapter. The first focuses on how our fellows are positioning themselves in relation to their imaginations of data science and the second describes students' experiences with navigating the data science field within our current social, political, and economic context.

### "It Seems More Like Number Crunching" – The Nuances of Imagining One's Self-Positioning

One theme that has consistently exhibited itself through the fellows' interviews involves the ways in which their imaginings of themselves as data scientists are intertwined with their competency beliefs, social networks, structural opportunities, navigational awareness, and past histories. Fellows described their perceptions of data science as an emerging field that is filled with both possibilities and uncertainties. This lack of clarity and stability was related to several fellows describing difficulty in imagining what life would be like as a data science professional. For example, one fellow, Carly (pseudonym) responded in this way when asked about what obstacles she perceives in becoming a data scientist:

> I don't really know much about data science. I'm not going to lie and how the field actually is, I know it's a bit of research. But to be honest, it seems more like number crunching. And, like examining numbers, so that's what it kind of gets, like, in my definition.

Like Carly, many fellows described their perception of data science, and their role in it, **in terms of the technical skills or competencies** they believed to be important (e.g., number crunching). It was helpful for our research team to be able to observe and contextualize these student perceptions – and their focus on technical skills (or competency beliefs) as a primary marker of their imagination of the field – in the outset of the program. A common observation in our analyses of student interviews, was how students focused on competency and technical skills in the absence of clear understandings of the other aspects that may go into becoming a data science professional.

### The Role of Prior Social Capital and Structural Opportunities

Many of the first-generation students in the program voiced initial difficulty in envisioning what life would be like as a data scientist or how data science could help them achieve their goals (which again, is not surprising for students beginning a program). However, our focus on understanding the identity trajectories of the students, enabled us to easily observe how these experiences from first-generation students were in juxtaposition to Gina (pseudonym), who shared that she came to the program already having prior research experience and is the daughter of a professor. When asked about what drew her to a data science program, Gina shared:

> I'm really excited to actually learn how to use R stuff, because I, like I said before, I've been in a research group for two and a half years, I understand that's something that they really need to use. And I'm excited because I know that'll be good for graduate school prep wise and stuff. I want to get a PhD in education.

Gina came into the program with both **social capital** (daughter of a professor, being comfortable in a research group already) and **structural opportunities** in the past (opportunity to be in a research group and observing what skills were important in that setting).

What was illuminating at this stage of the program and of learning about the students, was the **interaction** between the students' self-positioning (how they imagined themselves in relation to the field), their initial focus on technical skills as markers of their belonging, and how prior structural opportunities to be comfortable with technical markers (e.g. learning R) already gave some clarity for certain students (Gina) while other students experienced more vagueness and uncertainty. A key point here is that all the students in this specialized data science program were capable and interested in learning technical skills and tools, but they differed in how their competency beliefs and future imaginations intertwined in these crucial beginning stages of their learning experience.

### Self-Positioning and Future Choices

We also observed how students began with unclear and abstract notions of data science as a field, in addition to their anticipation that the field is fast-changing and unpredictable, made it difficult for students to position themselves or know how to see themselves as succeeding in the future. This perception compelled several students to describe **navigational knowledge** and **structural opportunities** as key areas that they were anxious about at the early stages of the program. For example, some students expressed a desire "to see data scientists working in the field and the obstacles that they have while working" and "getting experience doing specific work in the field" to gain practical experience for how to manage and navigate their career path.

Still, other learners in the data science cohort described how their different past experiences related to their perceptions of their own positioning in the field. This variation in backgrounds influenced the students' perception of their potential success, interest in the program, and their imagination as a future data scientist. For example, Emily (pseudonym) shared her worries about how her lack of past experiences related to her initial feelings of not belonging to her research group:

> … since, as I mentioned earlier, it's very new to me. I haven't had the opportunity to take public education classes. So a lot of topics that my lab is working on explaining can be all new to me. So sometimes I feel like I'm left like little behind from others in my lab group, so that sometimes, I kind of feel like maybe kind of don't belong in that group.

Here Emily is describing her perception that her lack of experience in public education puts her at a disadvantage in her research group – which focuses on studying urban students' learning experiences – because she requires more explanation to stay up to speed with her peers (who she perceives as already knowing tacit knowledge she does not have). These initial perceptions, whether accurate or not, contribute to the self-positioning narratives that are already forming in the beginning stages of her learning experience.

We offer a final example of how elements of past experiences, competency beliefs, and self-positioning come together in entirely different configurations for other learners. Charlotte was another data science fellow who came into the program with clearly defined research interests and research experience. Her past research experiences led her to have a more clearly defined interest in using data science to study issues of K-12 teacher retention. She was also motivated by her past experiences having a "revolving door" of math teachers in her own schooling. While these past experiences inspired Charlotte to have clarity in her purpose for joining the data science program, she also attributed her K-12 experience as giving her a weak background in mathematics (her own competency belief) which she worried might hinder her ability to thrive in the data science field.

It was illuminating to observe how her experiences provided both a positive inspiration for Charlotte's participation in the data science program, and simultaneously afforded a potential anxiety or obstacle in her imagination of herself in the field. Charlotte came into the program with clear research interests, but early in the program experienced difficulties in finding a faculty mentor on campus who specifically shared her interest in studying teacher retention. This challenge hindered her ability to join a research lab and find datasets to work with that exactly matched her well-formed research goals. From an identity lens, these developments meant that she perceived a lack of social connections, mentorship, and opportunities to practice her data science skills in personally relevant ways. As a consequence, she frequently voiced her worry that she may "run out of steam" and not persist in the program over time.

### *"That Pressure Impacts Some of My Decisions" – Financial and Social Mobility and Their Influence on Identity Stories*

A second theme in our initial analysis of fellows is that common career decisions about financial and social mobility can be understood in terms of the identity narratives learners are continuously crafting for themselves as they engage in new experiences. Several data science fellows expressed perceptions of financial investment and opportunity as key factors in their future imaginations of themselves as potential data scientists. Some fellows imagine data science as a possible safeguard for the financial investment of undergraduate education and as a set of skills and competencies that can help them achieve personal goals like supporting their family.

The rising costs of pursuing an undergraduate degree, specifically the cost of tuition and rising housing costs, were a major concern for many fellows. Some fellows shared in interviews that they commuted to campus from several hours away, took large loans to complete their degrees, or worked up to 30 hours a week in jobs such as food service, to afford the costs of their undergraduate education. For example, when asked about the challenges he currently faces navigating his academic and professional career path, Ben (pseudonym) shared:

> Mostly it's been in terms of like, being able to pay for school, it's been really, like rough trying to afford, like having to take out loans and whatnot to be able to afford to, like stay on campus and keep going to classes. Especially since, like, the interest rate has been so high, like the federal loans. It's like eight percent. It's kind of a crazy amount. And I'm like, Oh, my gosh.

The cost of a four-year institution is shocking to Ben, who prior to coming to this university was attending community college and living at home.

Some of the designed aspects of the data science program were well aligned to these potential challenges for students. For example, the stipends granted for participation in the fellowship made the program accessible for many students, both attracting them to the program and providing the financial support they needed to take a risk and spend time exploring data science. Financial concerns also often provided space for students to ask about navigational processes at the university institution. In the first days of the data science program, confusion about when and how the stipends would be distributed was of immediate concern for several students. At the conclusion of the first seminar and R workshop training, students asked many questions – not about R and learning skills with the statistical package – but about how the stipend may affect their financial aid and housing eligibility. These types of circumstances weigh heavily on how learners **navigate** their career paths and choose which opportunities to pursue in the short term and influence their long-term imaginations for where they want to go in their future.

Perceptions about social mobility and worth also played major roles in how students learn to navigate systems and position themselves for the future. For example, 12 of the 14 students in the program were pursuing humanities and social science degrees, traditionally seen as less valuable than STEM degrees. Many expressed worries about the value of their education and saw the data science program as a way to give themselves career flexibility, possibly mitigating the income disparity between STEM fields and their interests. Carly, whose desired major was phased out by the university, exemplifies the anxiety felt by learners unsure if their education and training will be worth it, saying:

> So that brought a lot of fear to me, like, hey, this school is getting rid of it. Are you calling my degree worthless? So I basically like, hey, I need to switch degrees. I don't know which one I just need to switch.

Fellows in our program often expressed their desire to seek data science related skills to better position themselves for an uncertain future. They recognized data as becoming increasingly important in many fields and the necessity to master data science practices to potentially achieve their goals. Learners in the program also shared how aspects of social capital and self-positioning related to their evolving sense of how they might persist in data science; particularly in sharing how they are not only navigating futures for themselves, but for the relationships they care about. Elena, for example, brought up how her first-generation status and the socioeconomic position of her family made her less likely to take big academic or career related risks:

> Taking care of my parents and like my younger siblings. So that's another pressure that maybe other students don't have. So that's why like, being first-generation, you have that pressure of like your family, and like, I'm the oldest child, so like, you need to succeed. That pressure impacts some of my decisions. And I feel like that could be a challenge in like the future as well. Maybe I'm not as seeking as I would be otherwise.

Elena's first-generation status and the responsibilities she has to her family influence what opportunities she seeks and how she navigates her future. Her **social capital** has a direct bearing on the kinds of **structural opportunities** that she sees as safe, valuable, and worth pursuing.

Similarly, Carly's future plans are driven by a desire to support her parents and help them get out of debt.

> So it's, like, stressful to, like, have all these like considerations like, location, income, is this job, you know kind of growing on the market is not growing on the market. And it's just like, those type of like, hesitancies that actually are kind of rooted in like, money, to be honest. That is, like, kind of, like, always gives me hesitancy when I try to like, figure out a career. Right. It's like, do something that has passion, but like, if I have passion, I can't really take care of people that I want to take care of.

The tension Carly experiences between personal interests she brings to the program, her nascent interest in data science, and her desire to provide for her family encapsulates the complex future imagining done by many learners today. Fellows come to data science within our current sociopolitical context, which offers a myriad of challenges. For example, we might understand the data science fellows' concerns and anxieties as direct reactions to rising income inequality (Horowitz, Igielnik, and Kochhar, 2020) and a consistent reduction of financial support for their higher education costs (Pew Charitable Trusts, 2019) in the United States. The pressures from this current context influence how they **navigate** their career pathways.

## Discussion and Implications

A central aim of this current chapter is to demonstrate the utility and potential for understanding data science education through a lens of identity. We see an identity lens as a potential way to fill in the gaps and expand our theoretical framing of the intersection of data science education and equity. For example, a common approach to designing for equity in a data science education program would provide learners with data and projects relevant to their interests and local community needs. These are important and vital pedagogical projects that could help learners begin to imagine STEM fields as more inclusive of their interests, as tools for community justice, and as agential activities that are more relevant to themselves as human beings (Clegg et al., 2020; Santo, Ahn, and Sarmiento, 2019). Conversely, data science programs might be designed from an access-perspective, with the goal to increase representation of diverse student populations. The design implications from this lens would prioritize the creation of programs that invite traditionally underserved learners and provide them opportunities to learn about a field but offer less insight about how these programs should be structured.

In our work we present an identity perspective as a complementary way to conceptualize and assess equitable outcomes in data science education. For example, pedagogical projects for equity in data science might be recognized as having substantial impacts on learners' imaginations for what is possible in STEM and their own role in it (*self-positioning* in our framework), help them grow in their *competency beliefs*, and foster deeper connections between their STEM activities and their social relationships to community and others (*social capital*). Programs that seek to increase representation of diverse student groups, on the other hand, could be recognized as ways to improve the *structural opportunities* that learners have access to, and perhaps give students new ways to *navigate* their pathways into a field such as data science. The identity lens we present here begins to illuminate the potential, and complementary, ways that different program designs are ideally suited to impact different facets of learners' development.

In addition, an identity lens also helps us to observe learners' experiences as they go through a program and highlight how program elements interact with these experiences. For example, the data science program in our study already had certain design elements such as providing students with a stipend and paying for students' GRE exams and graduate school applications. One might view these elements from the perspective of access as simply ways to increase representational diversity. However, we saw in our case analyses that economic considerations were deeply related to students' self-positioning in the field, their imaginations for the future, and how they decided to pursue or persist in future opportunities.

The data science program had other elements such as seminars and classes that taught key data science skills (e.g., using R to conduct research). These elements interacted with learners' identities in a variety of ways. For some learners who came in with prior structural opportunities, social capital, and competency beliefs, the courses amplified their pathways. For learners who came in with anxieties about their prior backgrounds and future possibilities, or less defined self-positioning, the courses strengthened their perception that technical skills were important for the field but may not inherently have helped to strengthen other aspects (e.g., social capital, navigation etc.). Conversely, program elements such as matching fellows with a faculty advisor or providing professional development seminars around navigating graduate school or data science as a field may touch on other aspects of learners' identity development. In our ongoing work, we aim to delve deeper into the myriad of student experiences as they progress through the data science program, and highlighting the variance in identity experiences and perceptions that they develop along the way.

One challenge that we recognize in this early work is that the difficulty of designing programs for identity development lies in the multifaceted ways that learners may perceive and interact with program elements. The issues of self-positioning and imagining the self in a career is not a one-step activity, but rather a developmental process that takes time to cultivate over many smaller experiences that accumulate and interact over time. Developing competency beliefs happens over time and necessitates careful reflection on self-progress and perceptions of the field. Expanding social capital, especially for marginalized learners, can be a difficult and non-linear process. Developing navigational understanding and pursuing relevant structural opportunities will be very different for learners at different stages of their educational or personal experiences. What is clear is that a data science education program that tends to the learner's identity development first must learn how learners position themselves in relation to data science. From this understanding of each learners' identity trajectory, data science programs could be structured for equity by engaging youth in scaffolded opportunities to reflect on their past, present, and future experiences; and continually seek to strengthen their imaginations for how data science can enrich their personal goals, life aspirations, and professional aims.

## References

Ahn, J. (2019). Drawing inspiration for learning experience design (LX) from diverse perspectives. *The Emerging Learning Design Journal*, 6(1), Article 1. https://digital-commons.montclair.edu/eldj/vol6/iss1/1.

Azevedo, F. S. (2011). Lines of practice: A practice-centered theory of interest relationships. *Cognition and Instruction*, 29(2), 147–184.

Bandura, A. (1997). *Self-efficacy: The exercise of control*. Freeman.

Bandura, A. (2001). Social cognitive theory: An agentic perspective. *Annual Review of Psychology*, 52(1), 1–26.

Bohnert, A., Fredricks, J., & Randall, E. (2010). Capturing unique dimensions of youth organized activity involvement: Theoretical and methodological considerations. *Review of Educational Research*, 80(4), 576–610.

Bourdieu, P. (2011). The forms of capital. (1986). *Cultural Theory: An Anthology, 1*, 81–93.

Brown, R., Cadena, N., Gatta, M., Ginsburg, N., Lee, M., Margolis, J., Nicholson, B., O'Neal, A., Pierson, R., Risteff, M., Shakur, A., & Smith, N. (2020). *Striving to Thriving. Occupational identity formation among Black and Hispanic young people and young people from households with lower incomes*. https://www.equitablefutures.org/wp-content/uploads/2020/10/Striving-to-Thriving-Full-Report-October-2020.pdf.

Calabrese Barton, A., Kang, H., Tan, E., O'Neill, T. B., Bautista-Guerra, J., & Brecklin, C. (2013). Crafting a future in science: Tracing middle school girls' identity work over time and space. *American Educational Research Journal*, 50(1), 37–75.

Calabrese Barton, A., Greenberg, D., Turner, C., Riter, D., Perez, M., Tasker, T., & Davis, E. A. (2021). Youth critical data practices in the COVID-19 multipandemic. *Aera Open, 7*, 23328584211041631.

Callahan, J., Ito, M., Campbell Rea, S., & Wortman, A. (2019). *Influences on occupational identity in adolescence: A review of research and programs*. Connected Learning Alliance.

Carlone, H. B., & Johnson, A. (2007). Understanding the science experiences of successful women of color: Science identity as an analytic lens. *Journal of Research in Science Teaching, 44*(8), 1187–1218.

Carpi, A., Ronan, D. M., Falconer, H. M., & Lents, N. H. (2017). Cultivating minority scientists: Undergraduate research increases self-efficacy and career ambitions for underrepresented students in STEM. *Journal of Research in Science Teaching, 54*(2), 169–194.

Clegg, T., Greene, D. M., Beard, N., & Brunson, J. (2020). Data everyday: Data literacy practices in a Division I college sports context. In *Proceedings of the 2020 CHI conference on human factors in computing systems* (pp. 1–13).

Clegg, T., Cleveland, K., Weight, E., Greene, D., & Elmqvist, N. (2022). Data everyday as community-driven science: Athletes' critical data literacy practices in collegiate sports contexts. *Journal of Research in Science Teaching.* https://doi.org/10.1002/tea.21842.

Davenport, T. H., & Patil, D. J. (July 15, 2022). Is data scientist still the sexiest job of the 21st century? *Harvard Business Review.* https://hbr.org/2022/07/is-data-scientist-still-the-sexiest-job-of-the-21st-century.

Eccles, J. S., & Wigfield, A. (2002). Motivational beliefs, values, and goals. *Annual Review of Psychology, 53*(1), 109–132.

Holland, D., Lachicotte, W. S. Jr, Skinner, D., & Cain, C. (2001). *Identity and agency in cultural worlds.* Harvard University Press.

Horowitz, J. M., Igielnik, R., & Kochhar, R. (2020). *Most Americans say there is too much economic inequality in the U.S., but fewer than half call it a top priority.* Pew Research Center. https://www.pewresearch.org/social-trends/2020/01/09/trends-in-income-and-wealth-inequality/

John, J. P., & Carnoy, M. (2019). The case of computer science education, employment, gender, and race/ethnicity in silicon valley, 1980–2015. *Journal of Education and Work, 32*(5), 421–435.

Jones, B. D., Tendhar, C., & Paretti, M. C. (2016). The effects of students' course perceptions on their domain identification, motivational beliefs, and goals. *Journal of Career Development, 43*(5), 383–397.

Kim, J. (2016). Who is teaching data: Meeting the demand for data professionals. *Journal of Education for Library and Information Science, 57*(2), 161–173.

Lave, J., & Wenger, E. (1991). *Situated learning: Legitimate peripheral participation.* Cambridge University Press.

Lang, C., Siemens, G., Wise, A., & Gasevic, D. (Eds.). (2017). *Handbook of learning analytics.* SOLAR, Society for Learning Analytics and Research.

Lee, V. R., Wilkerson, M. H., & Lanouette, K. (2021). A call for a humanistic stance toward K–12 data science education. *Educational Researcher, 50*(9), 664–672.

Li, D., & Koedel, C. (2017). Representation and salary gaps by race-ethnicity and gender at selective public universities. *Educational Researcher, 46*(7), 343–354.

Lopatto, D. (2007). Undergraduate research experiences support science career decisions and active learning. *CBE-Life Sciences Education, 6,* 297–306.

Markus, H., & Nurins, P. (1986). Possible selves. *American Psychologist, 41,* 954–969.

Martin, J. P., Stefl, S. K., Cain, L. W., & Pfirman, A. L. (2020). Understanding first-generation undergraduate engineering students' entry and persistence through social capital theory. *International Journal of STEM Education, 7*(1), 1–22.

McDonnell, L. M. (1995). Opportunity to learn as a research concept and a policy instrument. *Educational Evaluation and Policy Analysis, 17*(3), 305–322.

National Academies of Sciences, Engineering, and Medicine (2018). *Data science for undergraduates: Opportunities and options.* National Academies Press.

Oyserman, D., & Markus, H. R. (1990). Possible selves and delinquency. *Journal of Personality and Social Psychology, 59*(1), 112.

Persaud, A. (2021). Key competencies for big data analytics professions: A multimethod study. *Information Technology & People, 34*(1), 178–203.

Pew Charitable Trusts. (2019). *Two decades of change in federal and state higher education funding.* https://www.pewtrusts.org/en/research-and-analysis/issue-briefs/2019/10/two-decades-of-change-in-federal-and-state-higher-education-funding

Polman, J. L., & Miller, D. (2010). Changing stories: Trajectories of identification among african american youth in a science outreach apprenticeship. *American Educational Research Journal, 47*(4), 879–918.

Putnam, R. D. (2000). Bowling alone: America's declining social capital. In *Culture and politics* (pp. 223–234). Palgrave Macmillan.

Ryder, J., Leach, J., & Driver, R. (1999). Undergraduate science students' images of science. *Journal of Research in Science Teaching, 36*(2), 201–219.

Santo, R., Ahn, J., & Sarmiento, J. (2019). *Reclaiming digital futures: Lessons to help youth thrive through informal learning with technology.* University of California-Irvine. https://digitallearningpractices.org/wp-content/uploads/2019/03/Reclaiming_Digitial_Futures_Lessons_to_Help_Youth_Thrive_Through_Informal_Learning_with_Technology.pdf

Sfard, A., & Prusak, A. (2005). Telling identities: In search of an analytic tool for investigating learning as a culturally shaped activity. *Educational Researcher, 34*(4), 14–22.

Shepard, B., & Marshall, A. (1999). Possible selves mapping: Life-career exploration with young adolescents. *Canadian Journal of Counselling and Psychotherapy, 33*(1), 37–54.

Siemens, G., & Baker, R. S. D. (2012). Learning analytics and educational data mining: towards communication and collaboration. In *Proceedings of the 2nd international conference on learning analytics and knowledge* (pp. 252–254).

Takeuchi, L., Vaala, S., & Ahn, J. (2019). *Learning across boundaries: How parents and teachers are bridging Children's interests.* Joan Ganz Cooney Center at Sesame Workshop.

Yip, J., Clegg, T., Ahn, J., Bonsignore, E., Gubbels, M., Rhodes, E., & Lewittes, B. (2014). *The role of identity development within tensions in ownership of science learning.* International Society of the Learning Sciences.

# 9

# BUILDING THE INFRASTRUCTURE FOR QUANTITATIVE CRITICALISM IN RESEARCH METHODS COURSES

*Mario I. Suárez*

## Introduction

The explosion of artificial intelligence (AI) algorithms in everything from social media face filters to AI-generated texts has increased the need for data science education at all levels, from K-12 to higher education (Drozda, 2023; Levitt, 2022; Pek and Bauer, 2023; Provost and Fawcett, 2013). Past research has established that statistics courses (e.g., quantitative research methods courses) may have disproportionately negative outcomes for minoritized learners (Bateiha, Marchionda, and Autin, 2020; Opstad, 2020), particularly in introductory quantitative methods courses (Garfield et al., 2002). In part, this is attributed to outdated curricula that does not meet the needs of all learners (Aiken et al., 1990; Pek and Bauer, 2023; Schmidt, 1996). At the same time, there is evidence that top-tier journals in the field of education publish quantitative methods research that often utilizes intermediate and/or advanced statistical techniques (e.g., advanced regression models, factor analysis, path models; Hutchinson and Lovell, 2004). Thus, there is a need to not only make quantitative methods courses much more accessible to train the next generation of quantitative methodologists, but to make it relevant to the current societal needs so that it can be used for practical purposes.

While much attention with regards to introductory quantitative methods courses has been paid to pedagogy (e.g., Garfield et al., 2002; Liu, 2019), not much has been done with regards to updating the curriculum to address its racist and harmful history. Data science without a critical theoretical lens has historically resulted in harms to minoritized populations. For example, it has been used to support white supremacist ideas (Gillborn, Warmington, and Demack, 2018;

DOI: 10.4324/9781003364634-9

Strunk and Hoover, 2019) and to pathologize gender identity and sexuality (Keenan and Suárez, 2022). Though not new (Stage, 2007), critical quantitative methods have taken off in the last few years as a means to alleviate some of the harms that have been done by quantitative methods to minoritized communities and to bridge research with practice and policy (Castillo and Gillborn, 2022; Gillborn et al., 2018; Strunk and Hoover, 2019).

The purpose of this chapter is to present some of the ways that an introductory quantitative methods course and a K-12 middle school statistics unit can incorporate quantitative criticalism, as ways to build an infrastructure for quantitative criticalism across the K-16 statistics education spectrum. Part of the reasoning for including a K-12 statistics unit is that researchers have advocated to put as much effort into buy in for statistics K-12 courses as attention is paid to calculus pipeline (Garfield et al., 2002). Thus, K-12 students often fail to see how data-driven decision-making can be flawed and leave individuals out of analyses. I begin the chapter with an overview of quantitative criticalism and other theories that have stemmed from it, give some examples into some pitfalls of current takes on quantitative methods teaching, and provide background on two introductory quantitative methods courses. I then end by describing a middle school unit that can be introduced to students before they enter high school and higher education courses, so that the quantitative criticalist content and ideas can be slowly scaffolded.

*Positionality*

Before I begin describing what quantitative criticalism and other critical quantitative methods are, I would like to write a few things about my identity, and how my particular experiences have shaped the way I view quantitative methods. I am a Mexican American man of trans experience (i.e., a trans man), born and raised in the Texas-México border. As a first-generation college student who became a high school mathematics teacher at a Title 1 school, I saw first-hand how much of the effort my students put into their work often did not translate into upward mobility for them and their families. Thus, as I embarked on my doctoral journey, I knew I wanted to study how different interlocking systems of oppression played a role in the educational outcomes of minoritized students. Though I originally did not intend to be a quantitative methods researcher, I found advanced statistics accessible to me as a former Advanced Placement (AP) Statistics teacher.

## What Is Quantitative Criticalism?

Quantitative criticalism uses critical theory and quantitative methods to better understand the roles that power and privilege play in various outcomes

(Stage, 2007; Wells and Stage, 2015). Rather than viewing quantitative methods as fully objective, quantitative criticalists recognize that one's worldview and background play a role in the research design process, from drafting survey questions, to choosing the categories that participants respond to, to the interpretation of results. Quantitative criticalists also recognize the inherent challenges and tensions in highlighting certain populations as "reference" categories, for example, and silencing or rendering groups invisible by collapsing categories as a result of small sample sizes. All of these challenges have very real consequences, socially and politically, for minoritized individuals. Thus, as Stage (2007) writes:

> The critical researcher calls into question models, assumptions, and measures traditionally made under the positivist perspective. By using techniques such as interviews and observations, traditional critical researchers demonstrate situations and populations for whom the assumptions and models are fallacious. The critical quantitative researcher also questions models and assumptions but uses analysis of sociological and economic processes to demonstrate that for particular population groups, some widely accepted models and assumptions are inaccurate.
>
> *(pp. 9–10)*

## QuantCrit

There are other analytical lenses that quantitative criticalists have used to merge critical theory(ies) and quantitative methods. One such analytical lens is that of QuantCrit (Garcia, López, and Vélez, 2018; Gillborn et al., 2018), which merges critical race theory and quantitative methods. Gillborn et al. (2018) outline five principles that guide QuantCrit analyses:

1 The centrality of racism
2 Numbers are not neutral
3 Categories are neither "natural" nor given: for "race" read "racism"
4 Voice and insight: Data cannot "speak for itself"
5 Using numbers for social justice (p. 169)

That is, given the history of statistics as a tool of white supremacy (Noble, 2018; Zuberi, 2003; Zuberi and Bonilla-Silva, 2008), those who utilize Quant-Crit must recognize that numbers are not neutral nor "natural". Additionally, researchers give the data a "voice", which is informed by one's background and philosophical worldview. Thus, in order to alleviate some of the harms that numbers have created for people of color, it is important to utilize what we learn from quantitative studies for social justice.

## Why Learn about Quantitative Criticalist Methods?

Traditional methods of teaching quantitative courses fall short of addressing the needs of 21st century learners in a few ways. Not only do traditional methods not address its history of racism and harm, it also does not tend to move beyond traditional pedagogical styles. Some of these issues can relate to the levels of anxiety and reservations that graduate students have in taking introductory and much less, advanced, statistics courses (DeVaney, 2010).

### A Complex and Tumultuous Quantitative Methods History

Quantitative research methods courses fail to address the difficult histories of how statistics was used to create racial hierarchies where Black individuals were deemed inferior. While there are numerous examples, one concept that is prevalent in introductory statistics courses is that of linear regressions, used widely in statistics. However, we neglect to teach our students that one of the scientists widely known when addressing regressions, Francis Galton, pioneered eugenics methods of measuring achievement by measuring individual's physical attributes, to show Black people's intelligence as inferior to that of white individuals (Nobles et al., 2022; Zuberi, 2003). The use of racial stratification in statistics has undoubtedly contributed to what we now see in methods courses of using race as an independent or control variable, which renders it "fixed". James (2008) argues that doing so keeps social scientists from understanding the effects of racism as a system, rather than attributing a specific outcome to one's racial identification.

Another example in queer and trans studies field is that of John Money and Anke Ehrhardt's (1972) research of intersex bodies' operations to attempt to show evidence for Money's theory of gender malleability. Money was a psychologist and sexologist at Johns Hopkins University, where he carried out most of his research on intersex children. With cases where there were genetic twins and one of the children was intersex, Money and Ehrhardt advocated for what they called "corrective surgery", proposing that that child could be raised similarly to the non-intersex boy or girl and grow up "normal", thus supporting Money's theory of gender malleability. Using various *t*-tests types of analyses, they show in their book *Man & Woman, Boy & Girl: The Differentiation and Dimorphism of Gender Identity from Conception to Maturity,* that gender can be malleable, not taking into account the number of human rights violations on babies. One famous case, the case of the Reimer twins, where one of the children was operated on, ended up taking their own life as a result of the surgery and the aftermath that came from it. This particular example would be useful for students of quantitative methods to understand the implications of the work that we engage in, and how it has been used to harm individuals.

Strunk and Simpfenderfer (2022) have a special issue in press that bridges queer and trans theories with quantitative methods, titled "Queer Quantitative

Methodology in Educational Studies". Queer quantitative methods center queerness by disrupting normality and categories, among other things. When using secondary data analysis with big data, for example, queer quantitative methods problematize the categories in the datasets, which tend to be within a binary (for gender identity) or assume individuals are heterosexual and/or cisgender. Queer quantitative methods can also be used to center the experience of queer and trans individuals in education (Knowles et al., in press). That is, while most secondary data analysis tends to use cisgender and/or heterosexual individuals as reference categories when comparing groups, a queer quantitative lens could design studies that only collect data from queer and/or trans individuals (for examples, see James et al., 2016 or Suárez et al., 2022). This gives the message that queer and trans individual experiences are worthy of studying on their own, without having to compare to cisgender individuals as a reference category.

To conclude this section, I argue it is crucial for students of quantitative methods to engage with the history of the methods they are utilizing, especially the scientists behind those methods. In speaking about his book *Thicker Than Blood: How Racial Statistics Lie* (2003), Tufuku Zuberi mentions,

> I believe that social science is at its best when it is self-critical and relentlessly self-correcting. In order to be self-correcting we must be open to a critical evaluation of the methods we use and the conclusions that we have come to. In *Thicker Than Blood* I outline how statistical analysis was developed alongside a logic of racial reasoning. That the founder of statistical analysis also developed a theory of White supremacy is not an accident. The founders developed statistical analysis to explain the racial inferiority of colonial and second-class citizens in the new imperial era. I critically evaluate the history and practice of racial statistics to suggest ways in which social statisticians correct their practice.
>
> *Bonilla-Silva and Zuberi (2008, p. 5)*

In the next couple sections, I describe how I use quantitative criticalism to teach an introductory graduate-level statistics course, and how K-12 teachers could introduce some of these complex topics in a way that young learners can grasp. I have found success in reaching students who have expressed fear and/or anxiety in taking quantitative methods courses, and who end the course telling me that they learned not only the content, but that they appreciated my gentle and scaffolded approach to teaching statistics.

### Real-World Pedagogical Applications

It is well documented that traditional undergraduate and graduate statistics courses have not had pedagogical and curricular updates in a long time (Garfield et al., 2002; Hutchinson and Lovell, 2004). For example, a study of 201 introductory graduate quantitative methods courses in psychology in the United States and Canada by Aiken et al. (2008) revealed that statistical training is more prevalent and supported for those in laboratory settings, there is little to no entry-level coverage of advanced topics (e.g., multiple regression, structural equation models), measurement issues and newer techniques are not covered, among other findings. Aiken et al. (2008) completed a replication of an earlier study that had similar findings (Aiken et al., 1990). Studies with undergraduate statistics offerings at 373 institutions exposed a lack of introductory quantitative methods courses required at the undergraduate level, and when it is taught, is often separated into a statistics course and a methods course (Friedrich, Buday, and Kerr, 2000; Friedrich, Childress, and Cheng, 2018). Additionally, recent research by Oliver and McNeil (2021) showed that undergraduate data science programs tend to focus more on the mathematics and computer science of data science, but do not emphasize topics like statistical workflows and ethics of data use in their training. Oliver and McNeil (2021), along with others (Aiken, West, and Millsap, 2008; Friedrich et al., 2000; Friedrich et al., 2018) attribute the lack of emphasis on real-world connections to actual research-type activities to the training that those doing the instructing have. Together, these studies suggest that the anxiety that keeps graduate students from continuing advanced quantitative methods courses can be alleviated if there was a better emphasis at the undergraduate level (or below), along with pedagogical innovations in undergraduate-level statistics courses (Pan and Tang, 2004).

Quantitative criticalism is not a concept that needs to only be taught at the graduate level. There are ways of scaffolding some of the concepts at the K-12 levels through introductory statistics units and/or data science units. Doing so helps prepare students toward taking quantitative critical courses as undergraduates and thus, graduate students. In this section, I provide two case studies of how one could help build a foundation at the K-12 level through a middle school unit, and then present how I embed quantitative criticalism topics in introductory statistics courses I teach at the graduate level.

## K-12 Middle School Introduction to Statistics

Given the influx of information on the internet and as data science becomes more popular, educators are trying to seek out ways to introduce statistical concepts early on in students' K-12 education (Sagrans et al., 2022). However, available evidence suggests that teachers are not necessarily well-prepared to teach about data science concepts (Bowen and Roth, 2005). Thus, this need has paved the way for researchers to reconceptualize what K-12 mathematics and science curriculum should look like in a 21st century if data science is prioritized (Boaler and Levitt, 2019; LaMar and Boaler, 2021; Lee and Perret, 2022).

As a high school mathematics teacher at a Title 1 school, I always had a passion for making statistics accessible to all my students, not just my AP students. To that end, as doctoral students, a classmate (who was a middle school teacher herself) and I, created a lesson plan for teaching basic quantitative criticalist concepts in a middle school classroom that was published in Susan W. Woolley and Lee Airton's textbook, *Teaching about Gender Diversity: Teacher-Tested Lesson Plans for K-12 Classrooms* (Suárez and Wright, 2020). In the lesson, the teacher first hands out a 24-count box of crayons to each group of students and presents the class with a set of the three primary colors (i.e., red, yellow, blue) as well as a set of as many items from around the room as they would like. Then the teacher asks the class to place each item in a category that matches the crayons they were given. Students should note that there may not be exact crayon matches for all those items, in which case they either have to make the decision to not "count" that item or to categorize it under the closest match. They can create graphs (e.g., bar, circle) or other types of representations (e.g., table, fractions). The teacher can debrief with the class about the decisions that were made and why they were done in that way. They then do a second round and add more crayons, with as many subsequent rounds as time allows, increasing the number of crayons each round, along with the creation of numeric and graphic representations. Toward the end of the lesson or unit (if the teacher chooses to expand to more crayons), the students may start noticing that some items may need to be collapsed under specific crayon categories, but as more crayons are added, it may become easier to "match", though still, not every single item may be matched.

There are endless number of possibilities for this activity as an introduction to critical quantitative methods. Depending on the grade level, the teacher can add more complex statistical concepts, such as normality, survey categories, and descriptive statistics for older grades. Additionally, the teacher may have students examine different types of federal or large-study questionnaires such as the High School Longitudinal Study of 2009 (Duprey et al., 2018; Ingels et al., 2011, 2013) or the U.S. Trans Survey (James et al., 2016) and discuss implications of the categories for gender chosen for those surveys, for example. This could lead to robust and critical conversations about the implications of dichotomizing gender in early collections of the High School Longitudinal Study of 2009, compared to the diversity in responses for gender in the U.S. Trans Survey. For younger grades, such as elementary, the teacher may choose to not add as many crayon categories and not have as many items, as well as reduce the type of graphical and numeric representations. Rather than having class discussions about large-scale surveys, the teacher may have class discussions that lead to implications of categorization and "fit", among others.

### Graduate-Level Introductory Statistics

I have taught two introductory-level graduate courses in quantitative research methods during my time as a higher education instructor. Both are very similar in nature, though cater to different majors. One of the graduate courses is primarily students in the college of education but caters to master's students and audiology doctoral students, while the other caters to doctoral students in teacher education. Every single semester, without fail, I try to gauge students' level of comfort with statistics, and they are brutally honest—most say they just want to pass because it is a required course. Very few, if any of them in each class, mention an affinity for anything dealing with numbers, while the majority tell stories of always feeling afraid of mathematics. I reassure them that my goal is not just to have them pass, but that they will do well, and let them know that prior to teaching quantitative methods, I was a high school mathematics teacher who taught AP Statistics. While they believe me, I can tell that most of them are skeptical in their abilities to do more than just pass the course. This experience is not surprising, and well supported by research over a long period of time (e.g., Betz, 1978; Haynes, Mullins, and Stein, 2004; Li et al., 2021).

As a quantitative criticalist, my research and teaching tend to intersect. As such, the course begins with a few introductory readings on quantitative criticalism and QuantCrit (e.g., Gillborn et al., 2018; Stage, 2007) that tries to ease them into a different perspective of statistics than what they have in mind. I find that this eases their mind a bit. I provide a list of additional readings in case they are interested in pursuing more about quantitative criticalism. After the first two course meetings, we begin with a planning guide that is a basic introduction to statistics—from descriptive statistics and correlations, all the way to different tests of differences (e.g., $t$-tests, analyses of variance, multivariate analyses of variance), to basic regression and factor analyses. The content I teach is not much different than any basic textbook in statistics. However, the pedagogical strategies I implement may be somewhat different than what most graduate students are used to.

Each class meeting is structured in two parts: the guided lecture in the first half and the guided practice the second half. The first half of the course is likely similar to a typical lecture, though I pause every slide or two to ask students if they need clarification. We then continue on. As I guide them through a tutorial of the statistical software for the specific analysis for topic, I include scaffolded slides with the dataset we will use for the class. Each scaffolded slide contains instructions for the software, along with the output and a template interpretation. The template interpretation is color-coded in the same color highlight that corresponds to it from the output, so that the student knows where a specific number and/or keyword comes from in the output. For example, the $p$-value may be highlighted in red in the output and will correspond with specific terms in the template like "significant" and "$p < .001$". Once I set a specific color for one keyword (e.g., significant/significance), that will remain the same throughout the whole course unless I run out of colors, or if they are too similar to each other (e.g., magenta and red), which I will note to the class. After each tutorial is where quantitative criticalism comes into play. We address whether their specific hypotheses or research questions were answered, and discuss the research, practice, and political implications for the variables chosen, the categories used (especially if I had them practice collapsing categories), and how those choices made were done by a researcher who is human, informed by their worldview. Though that is not to say that the results are flawed, we discuss as a class how the decisions we make impact real people, often who are minoritized. The second half of the class meeting is a guided practice, where the students choose their own variables to craft hypotheses, research questions, and analyze the data for that statistical technique. It is guided, not only because I am available to troubleshoot and support them during that time, but because they are allowed to work in small groups of three or four students to support each other.

## Conclusion

This chapter presented the need to engage with quantitative criticalist methods in statistics courses. I provided an overview of how quantitative criticalism differs from traditional quantitative methods with examples. I then proceeded to provide examples of how I have used quantitative criticalism in my own quantitative research methods introductory courses, ending with how it could be implemented earlier on as a way to have students engage with these concepts prior to graduate school. Doing so, I believe, helps train students to think and understand statistics critically, and to understand the political and practical implications of their research.

## References

Aiken, L. S., West, S. G., & Millsap, R. E. (2008). Doctoral training in statistics, measurement, and methodology in psychology: Replication and extension of Aiken, West, Sechrest, and Reno's (1990) survey of PhD programs in North America. *American Psychologist, 63*(1), 32–50. https://doi.org/10.1037/0003-066X.63.1.32

Aiken, L. S., West, S. G., Sechrest, L., Reno, R. R., Roediger, H. L. III, Scarr, S., Kazdin, A. E., & Sherman, S. J. (1990). Graduate training in statistics, methodology, and measurement in psychology: A survey of PhD programs in North America. *American Psychologist, 45*(6), 721–734. https://doi.org/10.1037/0003-066X.45.6.721

Bateiha, S., Marchionda, H., & Autin, M. (2020). Teaching style and attitudes: A comparison of two collegiate introductory statistics classes. *Journal of Statistics Education, 28*(2), 154–164. https://doi.org/10.1080/10691898.2020.1765710

Betz, N. E. (1978). Prevalence, distribution, and correlates of math anxiety in college students. *Journal of Counseling Psychology, 25*, 441–448. https://doi.org/10.1037/0022-0167.25.5.441

Boaler, J., & Levitt, S. D. (2019). Modern high school math should be about data science—not Algebra 2. *Los Angeles Times.* https://www.latimes.com/opinion/story/2019-10-23/math-high-school-algebra-data-statistics

Bonilla-Silva, E., & Zuberi, T. (2008). Toward a definition of white logic, white methods. In T. Zuberi & E. Bonilla-Silva (Eds.), *White logic, white methods: Racism and methodology* (pp. 3–27). Rowman & Littlefield.

Bowen, G. M., & Roth, W.-M. (2005). Data and graph interpretation practices among preservice science teachers. *Journal of Research in Science Teaching, 42*(10), 1063–1088. https://doi.org/10.1002/tea.20086

Castillo, W., & Gillborn, D. (2022). *How to "QuantCrit:" Practices and questions for education data researchers and users. (EdWorkingPaper: 22-546).* https://doi.org/10.26300/V5KH-DD65

DeVaney, T. A. (2010). Anxiety and attitude of graduate students in on-campus vs. online statistics courses. *Journal of Statistics Education, 18*(1), 1–15. https://doi.org/10.1080/10691898.2010.11889472

Drozda, Z. (2023). Opinion: The world is changing fast. Students need data science instruction ASAP. *The Hechinger Report.* https://hechingerreport.org/opinion-the-world-is-changing-fast-students-need-data-science-instruction-asap/

Duprey, M. A., Pratt, D. J., Jewell, D. M., Cominole, M. B., Fritch, L. B., Ritchie, E. A., Rogers, J. E., Wescott, J. D., & Wilson, D. H. (2018). *High school longitudinal study of 2009 (HSLS:09) base-year to second follow-up data file documentation (NCES 2018-140).* National Center for Education Statistics, Institute of Education Sciences, U.S. Department of Education. https://nces.ed.gov/pubs2018/2018140.pdf

Friedrich, J., Buday, E., & Kerr, D. (2000). Statistical training in psychology: A national survey and commentary on undergraduate programs. *Teaching of Psychology, 27*(4), 248–257. https://doi.org/10.1207/S15328023TOP2704_02

Friedrich, J., Childress, J., & Cheng, D. (2018). Replicating a national survey on statistical training in undergraduate psychology programs: Are there "new statistics" in the new millennium? *Teaching of Psychology, 45*(4), 312–323. https://doi.org/10.1177/0098628318796414

Garcia, N. M., López, N., & Vélez, V. N. (2018). QuantCrit: Rectifying quantitative methods through critical race theory. *Race Ethnicity and Education, 21*(2), 149–157. https://doi.org/10.1080/13613324.2017.1377675

Garfield, J., Hogg, B., Schau, C., & Whittinghill, D. (2002). First courses in statistical science: The status of educational reform efforts. *Journal of Statistics Education, 10*(2), 1–14. https://doi.org/10.1080/10691898.2002.11910665

Gillborn, D., Warmington, P., & Demack, S. (2018). QuantCrit: Education, policy, 'Big Data' and principles for a critical race theory of statistics. *Race Ethnicity and Education, 21*(2), 158–179. https://doi.org/10.1080/13613324.2017.1377417

Haynes, A. F., Mullins, A. G., & Stein, B. S. (2004). Differential models for math anxiety in male and female college students. *Sociological Spectrum, 24*(3), 295–318. https://doi.org/10.1080/02732170490431304

Hutchinson, S. R., & Lovell, C. D. (2004). A review of methodological characteristics of research published in key journals in higher education: Implications for graduate research training. *Research in Higher Education, 45*(4), 383–403. https://doi.org/10.1023/B:RIHE.0000027392.94172.d2

Ingels, S. J., Pratt, D. J., Herget, D. R., Burns, L. J., Dever, J. A., Ottem, R., Rogers, J. E., Jin, Y., & Leinwand, S. (2011). *High school longitudinal study of 2009 (HSLS:09). Base-year data file documentation (NCES 2011-328).* U.S. Department of Education, National Center for Education Statistics. https://nces.ed.gov/surveys/hsls09/pdf/2011328_1.pdf

Ingels, S. J., Pratt, D. J., Herget, D. R., Dever, J. A., Fritch, L. B., Ottem, R., Rogers, J. E., Kitmitto, S., & Leinwand, S. (2013). *High school longitudinal study of 2009 (HSLS:09) base year to first follow-up data file documentation (NCES 2014- 361).* National Center for Education Statistics, Institute of Education Sciences, U.S. Department of Education. https://nces.ed.gov/pubs2014/2014361.pdf

James, A. (2008). Making sense of race in racial classification. In T. Zuberi & E. Bonilla-Silva (Eds.), *White logic, white methods: Racism and methodology* (pp. 31–45). Rowman & Littlefield.

James, S. E., Herman, J. L., Rankin, S., Keisling, M., Mottet, L., & Anafi, M. (2016). *The report of the 2015 U.S. Transgender survey.* National Center for Transgender Equality. https://transequality.org/sites/default/files/docs/usts/USTS-Full-Report-Dec17.pdf

Keenan, H. B., & Suárez, M. I. (2022). Toward trans studies informed theories and methods. In M. I. Suárez & M. M. Mangin (Eds.), *Trans studies in K-12 education: Creating an agenda for research and practice* (pp.13–36). Harvard Education Press.

Knowles, R. T., Hawkman, A. M., & Suárez, M. I. (in press). Making quant critical: Centralizing critical theories in quantitative social studies research. In XXXX (Ed.), *Rethinking research in social studies education* (pp. x–xx). Information Age Publishing.

LaMar, T., & Boaler, J. (2021). The importance and emergence of k-12 data science. *Phi Delta Kappan, 103*(1), 49–53. https://kappanonline.org/math-importance-emergence-k12-data-science-lamar-boaler

Lee, I., & Perret, B. (2022). Preparing high school teachers to integrate AI methods into STEM classrooms. *Proceedings of the AAAI Conference on Artificial Intelligence, 36*(11), 12783–12791. https://doi.org/10.1609/aaai.v36i11.21557

Levitt, S. D. (2022). Data science is the future. Let's start teaching it. *EducationWeek.* https://www.edweek.org/teaching-learning/opinion-data-science-is-the-future-lets-start-teaching-it/2022/01

Li, Q., Cho, H., Cosso, J., & Maeda, Y. (2021). Relations between students' mathematics anxiety and motivation to learn mathematics: A meta-analysis. *Educational Psychology Review, 33*(3), 1017–1049. https://doi.org/10.1007/s10648-020-09589-z

Liu, Y. (2019). Using reflections and questioning to engage and challenge online graduate learners in education. *RPTEL, 14*(3), 1–10. https://doi.org/10.1186/s41039-019-0098-z

Noble, S. U. (2018). *Algorithms of oppression: How search engines reinforce racism.* NYU Press. https://nyupress.org/9781479837243/algorithms-of-oppression/

Nobles, M., Womack, C., Wonkam, A., & Wathuti, E. (2022). Science must overcome its racist legacy: Nature's guest editors speak. *Nature.* https://www.nature.com/articles/d41586-022-01527-z

Oliver, J. C., & McNeil, T. (2021). Undergraduate data science degrees emphasize computer science and statistics but fall short in ethics training and domain-specific context. *PeerJ Computer Science, 7*, e441. https://doi.org/10.7717/peerj-cs.441

Opstad, L. (2020). Attitudes towards statistics among business students: Do gender, mathematical skills and personal traits matter? *Sustainability, 12*(15), Article 15. https://doi.org/10.3390/su12156104

Pek, J., & Bauer, D. J. (2023). How can we move advanced methodology into practice more effectively? *Policy Insights from the Behavioral and Brain Sciences, 10*(1), 3–10. https://doi.org/10.1177/23727322221144649

Provost, F., & Fawcett, T. (2013). Data science and its relationship to big data and data-driven decision making. *Big Data, 1*(1), 51–59. https://doi.org/10.1089/big.2013.1508

Sagrans, J., Mokros, J., Voyer, C., & Harvey, M. (2022). Data science meets science teaching. *The Science Teacher, 89*(3), 64–69. https://www.nsta.org/science-teacher/science-teacher-januaryfebruary-2022/data-science-meets-science-teaching

Schmidt, F. L. (1996). Statistical significance testing and cumulative knowledge in psychology: Implications for training of researchers. *Psychological Methods, 1*(2), 115–129. https://doi.org/10.1037/1082-989X.1.2.115

Stage, F. K. (2007). Answering critical questions using quantitative data. *New Directions for Institutional Research* (133), 5–16. https://doi.org/10.1002/ir.200

Strunk, K. K., & Hoover, P. D. (2019). Quantitative methods for social justice and equity: Theoretical and practical considerations. In K. K. Strunk & P. D. Hoover (Eds.), *Research methods for social justice and equity in education* (pp. 191–201). Palgrave Macmillan. https://link.springer.com/chapter/10.1007%2F978-3-030-05900-2_16

Strunk, K. K., & Simpfenderfer, A. D. (2022). *Queer quantitative methodology in educational studies (Call for papers)*. https://www.dropbox.com/s/ejxbu95uvibllce/Call%20for%20Papers%20052022.pdf?dl=0

Suárez, M. I., McQuillan, M. T., Keenan, H. B., & Iskander, L. (2022). Differences in trans employees' and students' school experiences. *Educational Researcher, 51*(5), 352–358. https://doi.org/10.3102/0013189X221100834

Suárez, M. I., & Wright, K. B. (2020). A critical approach to teaching data management and analysis categories. In S. W. Woolley & L. Airton (Eds.), *Teaching about gender diversity: Teacher-tested lesson plans for k-12 classrooms* (pp. 139–148). Canadian Scholars Press.

Wells, R. S., & Stage, F. K. (2015). Past, present, and future of critical quantitative research in higher education. *New Directions for Institutional Research, 2014*(163), 103–112. https://doi.org/10.1002/ir.20089

Zuberi, T. (2003). *Thicker than blood: How racial statistics lie*. University of Minnesota Press.

Zuberi, T., & Bonilla-Silva, E. (2008). *White logic, white methods: Racism and methodology*. Rowman & Littlefield.

# 10

# CLOSING THOUGHTS AND FUTURE DIRECTIONS

*Colby Tofel-Grehl and Emmanuel Schanzer*

The term data science is used in widely varied ways across academic and societal contexts; this variety of meaning and nuance can make the field broadly ill-defined and lead to confusion about how data science differs from both computing and statistics (Provost and Fawcett, 2013). Further complicating the issue is the ways that historically marginalized groups tend to be ill-served or absented from consideration within traditional data science curricula and classes.

Today's children will face unexpected challenges and opportunities growing up and living in a data-centric world. As they move through life, their engagement with the world is recorded through a series of transactions, mouse clicks, and other microlevel decisions that generate information about them. Similarly, the world offers them localized and global information in a myriad of ways that require them to sift, distill, and synthesize information to inform their decision making. Who they are seen to be, who they see themselves as, and what opportunities they have will all be shaped by the data they engage and the ways in which they engage it. When digital technology first became omnipresent, we began to differentiate between youth who were so-called digital natives, using technology from childhood, and those who came before who learned how to engage with it as adults. Much like Gen Z never knew a time without technology, they will never know a time when the quantification and commodification of their data was not part of their lived reality.

But algorithms err. Despite being programmed to elucidate, they often obfuscate, leaving swaths of our communities absented and unrepresented in the "reality" they offer up as truth. For example, in 2019, researchers found the commonly used algorithm that determines care priority within U.S. hospitals

DOI: 10.4324/9781003364634-10

demonstrated extreme racial bias such that Black patients needed to be sub-stantially more sick than their white counterparts to be deemed worthy of needing the same level of care (Obermeyer et al., 2019). Data—and by exten-sion data science as a field—is rife with misrepresentation, misunderstanding, and misuse. Claims of absoluteness are a mirage. What data and data science offer us are ways of looking at and questioning various phenomena through the lens of the data scientist. Engaging in equitable data science work requires us to ask:

*Whose story or perspective is centered in this data/analysis?*

*Whose story or perspective is absented from this data/analysis?*

*What intent and purpose drives this work?*

*What or who might be missing?*

Engaging these questions raises one more: Who are the appropriate people to do this work? A concern sometimes voiced regarding equity work is that white scholars may not be best suited to the work, both because of their positionality and the potential for them to draw focus away from scholars of Color. While these are important points to consider, the alternative is that a portion of the scholarly community abdicate our shared responsibility to make society more equitable and, by default, place the burden of this work solely on members of communities that have been harmed by inequitable systems and structures of oppression. Accordingly, it is our view that we must all seek to maximize equity as we engage across the broad field of data science and data science education, actively seeking opportunities to learn from the experiences and insights offered by deep scholarship in equity. Scholars of diversity, equity, and inclusion offer a set of lens, frameworks, and rich experiences through which all scholarly efforts in schools and communities should be informed. From there, the work and the responsibility for engaging it belongs to all of us. In order for equity work to be fully integrated and centered within classrooms and learning spaces, all work needs to bring an equity lens to the conversation. Simply saying that equity is not what one does, when working with educators and students, is insufficient. If, as authors within this book have established, all data and data science is inher-ently political, shaped from its conception through collection and analysis by the data scientist, then data science can never be free of the temporal and social context in which it exists. Data science education research therefore requires *everyone*, regardless of comfort or expertise, to elucidate the structures and fac-tors that influence, impact, and inform work with teachers and students.

The authors within this volume have done an admirable job laying out the historic evolution of data science education (e.g., Walker et al.), emerging frameworks for conceptualizing learners' relationships to data science (e.g., Lee), and the essential need for rich, diverse opportunities for identity development in data science (Ahn et al.). Further, they offer exemplars of some of the varied ways equity can be centered and central to data science education research (e.g., Grover et al.; Shriener et al.; Suarez). Mixing case examples and theoretical expositions, we hope to have opened a door to the possible and begun a dialogue that needs to be engaged by a wide range of communities and stakeholders.

In the opening chapter, Schanzer argues that data science is an approach to knowing and understanding that is more than merely a sub field within statistics or computer science. Instead of relegating data science to a footnote within a larger math or computing pathway, he argues that the diverse and interdisciplinary nature of the field requires that it be considered across multiple contexts in order to be equitably and responsibly taught at all: core competencies in mathematics and computing are necessary, but not sufficient. Instead, he argues for the applied development of concrete data science literacies around data safety, comprehension, analysis, and interpretation across the curriculum. By weaving data science into the subjects that already reach every child, he argues that we can fully recognize the critical literacies of data science that youth need both for their personal success but also for their successful participation within society. If the push for data science education is to be equitable, he writes, "data science must legitimize and elevate the critical questions raised in humanities classes – not be the next movement to chip away at them."

Walker and colleagues then offer a historic perspective on the transition from computer science to data science models and the deep, transformative nature of culturally responsive computing. They lay the groundwork for developing our understanding of how culturally responsive teaching would both be conceptualized and ultimately operationalized within a data science context. Centering a participatory model and youth agency, Walker and colleagues explicate a model of culturally responsive data science that grows out of well-established work within computer science.

Hansen et al. offer a case example of a data science curricular unit designed to scaffold learning about the experiences and realities of Indigenous communities in the western United States. By examining the shrinking lands of the Black Feet nation from their historic lands in the 1800s and tracking the changes enacted through various treaties, youth are invited into a world of windows and mirrors (Bishop, 1990) to understand, appreciate, and contextualize the experiences of the Black Feet Nation. When taught within a white community, as done in their case example, the project offers a window of understanding for white youth as to real-world implication of the treaties, brokered and broken, by the U.S. government. This data visualization leads youth to ask important questions about the longstanding impacts of these broken promises and explicates the intergenerational challenges created by them. As part of this research, Hansen et al. examined the ways that youth engaged with personal and distal data as they made meaning of data. As youth engaged with personal and group level data local to themselves, they were able to begin to develop their own understandings of what data is and how to make sense of it. Then, later on, working with distal data, youth began to apply their newly developed skills to ask questions about the impacts, immediate and long term, to the Black Feet Nation of ever-changing land boundaries. By exploring this data and asking questions, youth were able to better understand the forced social changes on nomadic peoples' hunting and food procurement approaches when restricted to minimal lands. This window into the experiences and challenges experienced by others allowed students to appreciate the nuanced problems associated with the U.S. government's broken promises to the Indigenous peoples of Montana.

Grover et al. describe a newly developed set of integrated curriculum, pedagogy, and data science tools to support and scaffold girls' interest and critical consumption of data and data science. Beginning with tools that can work flexibly with diverse datasets to facilitate rapid transition from accessing data to analyzing it, the curriculum draws students' attention to the ways in which bias can manifest within data and its analysis. They further position data science projects as collaborative endeavors to support engagement, drawing in students with differing levels of skill and prior exposure. Through preliminary data, they demonstrate the promise of an integrated approach that bundles tools, content, and scaffolded engagement strategies to help students who may not be likely to participate in computing and data science projects understand and leverage the power of sophisticated data analysis with low barriers to entry.

In Tofel-Grehl et al., we are offered a slightly modified approach to equitable data science education. Authors engage in reflection on their limited ability as white people to be culturally responsive. In response to this tension, they offer us another way to frame and develop equitable data science education through community centering. In centering the work around community, the authors give prominence and intentional centrality to the worldviews of folks with whom they work; they focus attention on folks current needs and identities, rather than engaging a single instantiation or conception of culture for a community of which they are not members. This offers a different way for white scholars to attune to issues of equity without imposing their own definitions of responsiveness to culture. By centering community and working to create spaces of youth rightful presence (cf. Calabrese Barton and Tan, 2019), the authors showcase a model of community centered data science. Youth within this project engaged in a community centered approach to data science that mapped onto the science practices laid out in the NGSS.

Lee dives deeply into a case of how one can design preservice teacher education to be data science rich. In understanding the needs of preservice educators, Lee explores students' perceptions of the pervasiveness of data and how we, as teacher educators, can scaffold learning and improve outcomes around who engages in and discusses data science within their teaching spaces. Of most note, this chapter engages in a rich discussion of the intentional design process undertaken to create an equity centered space for exploring data and data science. With scaffolded intent, he is able to weave together a space of engagement and reflection for future teachers that centers methods and perspectives of equity within the higher education setting.

Ahn and colleagues focus on bringing forward a call to action for data science educators. They argue for the intentional scaffolded development of data science identities within and across communities to ensure the development of a diverse data science workforce. Making meaning both within their class and across the field, their work is rooted in understanding and fostering meaningful and contextualized data science identities in higher education. Of particular interest is the model their university adopted to facilitate this identity development through an intentional cohort model attending to the real-world concerns of stakeholders by providing fiscal support and mentoring to navigate programs, structures, and disciplinary pathways related to data science learning. The fully supported model they offer allows for rich and sustained data science identity development for participants. How they broaden participation and recruit groups currently not self-selecting into their program remains to be seen, but the intentional and transparent effort being made signals tremendous promise for offering an applicable and widely deployable model for scaffolding data science identity.

A great wealth of the work done to date in data science education centers STEM disciplinary content or classes. Shriner and Guzdial offer an examination of the development of data science teachers outside of STEM. Working with social studies teachers, their chapter explores the tools and pedagogies of scaffolding social studies teacher knowledge and instructional approaches for asking data science based questions. They provide a rich case of the ways that data science education can be the rising tide that lifts all proverbial boats. By engaging teachers in developing and using tools that address their instructional content goals while answering data science and visualization questions, they offer a glimpse at a technologically rich return to civics education. With civics arguably the most important educational competency within public education, their approach demonstrates the rich range of opportunities for data science education to develop a more informed and critically aware populous.

While we argue that data science is a distinct space from statistics, the two often occupy common ground. Everyday statistics education takes for granted the structures and assumptions on which it is built. Understanding the ways that these assumptions disengage and marginalize groups is essential to acknowledging and addressing the power differentials that shape engagement. Suarez offers an overview and exemplars of quantitative criticalism in action within the statistics classroom. Engaging a critical lens within data science education affords learners and educators the opportunity to explore the structures and underlying perspective biases that obfuscate "big T" truth within data analysis and interpretation. By engaging quantitative critical lenses within data science education, Suarez explicates the ways that he is able to recenter the absented stories and nuanced perspectives within the statistics classroom, regardless of student age.

Each of the case examples discussed in this volume recenters conversations. They seek to engage learners in understanding the stories historically left untold or marginalized. Whether constructing data visualizations of the impacts changing treaties had on Indigenous people, as Hansen and colleagues do, or exploring the ways that secondary social studies teachers can develop their skills and understandings of data science within their content area as laid out by Shriner and Guzdial, each chapter serves to elucidate the ways that data science scholars and educators can *and must* center humanism (Lee, Wilkerson, and Lanouette, 2021) and rightful presence (Calabrese Barton and Tan, 2019) to ensure our understandings and experiences with data reflect the complex histories and realities of the messy, and oft ugly, reality in which we all exist. As Ahn and colleagues remind us, identity is not independent of data or data science. It is integral in defining of measures, the assumptions used in analysis, and the interpretation of results. Further, it is not independent of the ways in which people learn these or any other skills.

This volume by no means offers a single or definitive approach to fostering equity in data science education. Much to the contrary, we see in this volume the perspectives, epistemologies, and approaches of the chapter authors. Our hope as editors has been to begin a conversation, generate ideas, and learn in partnership with chapter authors. Specifically, we hope to go beyond epistemologies of data science to begin consideration of axiology as well.

Axiology frames the values with which we infuse bodies of scholarly research and practice. For example, some of the work presented in this volume adopts values explicit in the learning sciences such as a heightened emphasis on the role of context as a foundational for learning and the valuing of co-design as a legitimation of learners' funds of knowledge. We feel it is essential to infuse data science with a valuing of not just persons, but *personhood*—the rightful presence of a person's full concept of themselves as salient and necessary to contribute, evaluate, or utilize data. Lee, Wilkerson, and Lanouette (2021) draw attention to the importance of a humanistic stance toward data science education in which personal experience with data, cultural tools and data practices, and political narratives are salient elements of learning experiences. We extend these valued types of data engagement by recognizing that these are necessary components of human experience and personal identity. Without permitting people the agency to bring the totality of themselves to the practices of data science—that is, to be rightfully present and empowered to manifest who they are within data construction, analysis, interpretation, and use—we bar people from fully equitable participation in data science and the benefits that society can derive from it.

Data science is as fallible as the people involved in the generation, collection, and analysis of data. There will undoubtedly be others who take different stances, bring different lenses, and hold different ideas about the path forward. We trust in their perspectives, hope for dialogue, and look forward to learning from them as well. In all things, we learn from doing. We hope this work offers a space of learning and consideration for the work ahead. This book, we hope, serves as a starting point for a conversation around what equitable data science might look like right now within various K-16 spaces.

## References

Bishop, R. S. (1990, March). Windows and mirrors: Children's books and parallel cultures. In *California State University reading conference: 14th annual conference proceedings* (pp. 3–12).

Calabrese Barton, A., & Tan, E. (2019). Designing for rightful presence in STEM: The role of making present practices. *Journal of the Learning Sciences*, *28*(4–5), 616–658.

Lee, V. R., Wilkerson, M. H., & Lanouette, K. (2021). A call for a humanistic stance toward K–12 data science education. *Educational Researcher*, *50*(9), 664–672.

Obermeyer, Z., Powers, B., Vogeli, C., & Mullainathan, S. (2019). Dissecting racial bias in an algorithm used to manage the health of populations. *Science*, *366*(6464), 447–453.

Provost, F., & Fawcett, T. (2013). Data science and its relationship to big data and data-driven decision making. *Big Data*, *1*(1), 51–59.

# INDEX

Note: Page numbers in *italics* represent figures and in **bold** indicate tables.